Human Models of Central Sensitization Assessed by Nociceptive Withdrawal Reflexes and Reflex Receptive Fields

Human Models of Central Sensitization Assessed by Nociceptive Withdrawal Reflexes and Reflex Receptive Fields

PhD Thesis by

José A. Biurrun Manresa

*Integrative Neuroscience Group,
Center for Sensory-Motor Interaction,
Department of Health Science and Technology,
Aalborg University, Denmark*

River Publishers

Aalborg

ISBN 978-87-92329-56-1 (paperback)
ISBN 978-87-92329-95-0 (e-book)

Published, sold and distributed by:
River Publishers
P.O. Box 1657
Algade 42
9000 Aalborg
Denmark

Tel.: +45369953197
www.riverpublishers.com

Table of contents

Preface

This Ph.D. thesis is the result of work carried out between October 2007 and October 2010 at the Center for Sensory-Motor Interaction, Aalborg University (Denmark), supported by The Danish Research Council for Technology and Production Sciences (FTP). Five months of this period, between September 2009 and February 2010, were carried out at the University Hospital of Bern, Inselspital (Switzerland) as part of an ongoing collaboration between these two institutions. This stay abroad was supported through an EliteForsk travel stipend, granted by the Danish Ministry of Science, Technology and Innovation.

This Ph.D. dissertation is a contribution to the understanding of the mechanisms underlying central sensitization of spinal nociception in humans. The aims of this Ph.D. project were to explore different models of central sensitization in humans and to assess them objectively using nociceptive withdrawal reflexes and reflex receptive fields.

The thesis contains four chapters. The first chapter presents the necessary background knowledge on central sensitization, the aim of the project and an overview of the dissertation. The second chapter depicts the methodology used for objective assessment of central sensitization, using nociceptive withdrawal reflexes and reflex receptive fields. The third chapter describes the human models for central sensitization studied during this project, and the thesis is completed with a fourth chapter with a brief conclusion and future perspectives.

The core of this dissertation is based on four original papers that were either published or submitted to international peer-reviewed journals. In addition, two peer-reviewed journal papers, two peer-reviewed conference papers and several conference abstracts complement the scientific work conducted in this project.

Acknowledgments

This Ph.D. thesis has only been possible thanks to the help and support of many people to whom I am deeply grateful. In particular, I would like to express my most sincere gratitude to Ole K. Andersen, because I could not have expected or imagined a better professional and personal relationship with my supervisor before coming to Denmark, in addition to the great scientific environment he provided. I deeply appreciate his constant guidance and support, his confidence in me and his patience to bear with my strange working schedule. I am also most grateful to Carsten D. Mørch for his excellent predisposition to help me at all times (especially when Ole was away) and for all the fruitful scientific discussions, valuable comments and suggestions, and his generous assistance in so many things (as for example, the Danish translation of the summary in this thesis). Thanks to John Hansen as well, for sharing with me his passion for teaching and technology.

I would also like to thank all the people who helped me with the studies that constitute this thesis, from those who discussed hypothesis and ideas with me, to those who participated through long (and painful) hours during my experiments. I would especially like to acknowledge Michele Curatolo, Alban Neziri and Martina Dickenmann for their kindness and help during our stay in Bern, Emanuel van der Broeke and Oliver Wilder Smith for my short but very pleasant stay in Nijmegen, Michel B. Jensen for all the hours we shared planning, discussing and carrying out experiments and Lars Arendt-Nielsen for his insightful reviews of our papers. I also appreciate the kindness and good predisposition of all the administrative and technical and staff at SMI, especially from Susanne Nielsen, Lone S. Andersen, Birgitte K. Hansen, Peter Thonning and Jan Stavnshøj.

My sincere gratitude to all my friends here at SMI: Kersting Jung, Sara Finocchietti, Carolina Vila-chã, Silvia Muceli, Rogerio Hirata, Dejan Ristic, Jonas Emborg, Thomas N. Nielsen, Thomas Lorrain, Afshin Samani, Francesco Negro, Héctor Caltenco, Steffen Frahm, Mogens Nielsen, Shilpa Razdan, Jovana Kojovic and all the others that are elusive to my poor memory: I am really fortunate to have spent so many delightful hours amongst you, that certainly made work and life in Denmark an incredible experience. In this regard, I am deeply in debt with my Argentinian friends in Aalborg, who were truly a family for me during all this time and especially when I was just arrived, alone and far away from all the things that I cherished. It was your kindness, affection and encouragement that helped me through those difficult periods of transition during the first months here.

I also wish to express my gratitude and appreciation to my family in Argentina: my mom Adriana, my Dad José, my sister Jimena and my brother Juan Manuel for always managing to stay close to my heart despite the distance. Finally, my deepest love to my wife Marta and our son Matías, who give meaning to everything in my life. This thesis is dedicated to them.

English summary

Central sensitization is believed to be one of the key mechanisms that are responsible for many of the temporal, spatial and threshold changes in pain sensitivity in acute and chronic clinical pain settings. Uncovering the mechanisms that initiate and maintain central sensitization is of utmost importance in order to develop more effective treatments against painful conditions. However, clinical trials in pain patients are usually a costly and time-consuming process, and they always involve a degree of heterogeneity in regards to the factors that could potentially interact with the mechanisms under evaluation. Thus, prior evaluation of the efficacy of new drugs or alternative methods for pain relief in human surrogate models of central sensitization in healthy volunteers may serve as an initial proof of concept and may also help improving study designs and defining relevant efficacy parameters in subsequent clinical trials.

Within this context, the aims of this Ph.D. project were to explore different models of central sensitization in humans and to assess these models objectively using nociceptive withdrawal reflexes (NWR) and reflex receptive fields (RRF). To this end, four studies were carried out, referred to as Study I to IV. In Studies I and II, a number of methodological aspects about the NWR and the RRF were addressed in healthy volunteers and chronic pain patients, in order to find the best parameters for NWR and RRF quantification in relation to spinal nociception. In Studies III and IV, the NWR and the RRF were used as objective assessment methods for human surrogate models of central sensitization. Such models were based on conditioning electrical stimulation on the skin and chemical irritation induced by intramuscular injection of capsaicin.

The results from Studies I and II indicated that the NWR and the RRF are robust and reliable measures of spinal nociception in healthy volunteers as well as in chronic pain patients. Moreover, the results from Study III and IV showed that central sensitization models could be established using two different types of nociceptive activation, and the outcome of these models was successfully assessed using the NWR and the RRF. In conclusion, the NWR and RRF are valid alternatives for objective assessment in experimental and clinical pain research towards a better understanding of the mechanisms behind acute and chronic pain conditions.

Dansk sammenfatning

Central sensibilisering menes at være en af de vigtigste mekanismer, der er ansvarlige for mange af de tidslige, rumlige og tærskel ændringer under smerte sensibilitet i akutte sammenhænge og kroniske smerteklinikker. At afdække de mekanismer, der starter og opretholder central sensibilisering er yderst vigtigt med henblik på at udvikle mere effektive behandlinger mod smerte. Men kliniske smerteforsøg i er normalt en dyr og tidskrævende proces. Forudgående vurdering af effekten af nye lægemidler eller alternative metoder til smertelindring i menneskelige surrogatmodeller af central sensibilisering hos raske forsøgspersoner kan derfor tjene som en første "proof of concept", og kan også bidrage til at forbedre studiedesign og definere relevante effektparametre i de efterfølgende kliniske forsøg.

I denne forbindelse sigtede dette Ph.D. projektet på at udvikle pålidelige modeller for central sensibilisering hos mennesker og til at vurdere disse modeller objektivt ved hjælp af nociceptive afværgereflekser (NWR) og refleks-receptive felter (RRF). Til dette formål, blev fire undersøgelser, der omtales som Studie I-IV, udført. I Studie I og II blev en række metodiske aspekter af NWR og RRF behandlet i raske frivillige forsøgspersoner og i kroniske smertepatienter, for at finde de bedste parametre til at kvantificere NWR og RRF i forbindelse med spinal nociception. I Studie III og IV, blev NWR og RRF testet som objektive metoder til vurdering af humane surrogatmodeller af central sensibilisering. Disse modeller er baseret på konditionerende elektrisk stimulation på huden og kemiske irritation fremkaldt ved intramuskulær injektion af capsaicin.

Resultaterne fra Studie I og II indikerede at NWR og RRF er robuste og pålidelige mål for spinal nociception hos raske forsøgspersoner såvel som hos kroniske smertepatienter. Desuden viste resultaterne fra Studie III og IV viste, at central sensibilisering modeller kan etableres ved hjælp af to forskellige typer af nociceptive aktivering, og at effekten af disse modeller kunne vurderes ved hjælp af NWR og RRF. Konklusionen er, at NWR og RRF er brugbare alternativer for objektiv vurdering i eksperimentel og klinisk smerteforskning sigtende på en bedre forståelse af mekanismerne bag akutte og kroniske smertetilstande.

List of abbreviations

ASI	adjusted stimulation intensity
CR	coefficient of repeatability
CV	coefficient of variation
EMG	electromyography
EP-T	electrical pain threshold
FSI	fixed stimulation intensity
HFS	high-frequency stimulation
ICC	intraclass correlation coefficient
LA	limits of agreement
LFS	low-frequency stimulation
NI	non-injured
NMDA	N-methyl-D-aspartate
NWR	nociceptive withdrawal reflex
NWR-T	nociceptive withdrawal reflex threshold
QST	quantitative sensory test
RMS	root-mean-square
ROC	receiver operating characteristic
RRF	reflex receptive field
SCI	spinal cord injured
SEM	standard error of the mean
SOL	soleus
TA	tibialis anterior
TKEO	Teager-Kaiser energy operator
WDR	wide-dynamic-range

Chapter 1.

Introduction

Long-lasting, activity-dependent synaptic plasticity in the nociceptive system was first documented by Woolf in 1983; the description corresponding to an immediate-onset increase in the excitability of neurons in the dorsal horn of the spinal cord after brief, intense nociceptive input. Since this effect could not be caused solely by peripheral mechanisms [18,20], this phenomenon was initially termed *central sensitization*. Nowadays, this concept describes an enhanced responsiveness of nociceptive neurons in the central nervous system to their normal and/or sub-threshold afferent input [14], as well as the enlargement of neuronal receptive fields [2,13]. Therefore, most of the forms of synaptic plasticity that occur in the spinal cord in response to noxious stimuli, from short-term effects that only persist for a few seconds like wind-up [6], to more long-lasting phenomena, such as activity-dependent central sensitization [18] and spinal long-term potentiation [19], are encompassed into this wider definition [7].

Over the years, increasing evidence has been found linking central sensitization with pathological pain states. Indeed, central sensitization is responsible for many of the temporal, spatial and threshold changes in pain sensitivity in acute and chronic clinical pain settings, exemplifying the fundamental contribution of the central nervous system to the generation of pain hypersensitivity [13]. Therefore, uncovering the mechanisms that initiate central sensitization is of utmost importance in order to develop more effective treatments against painful conditions. However, regulatory guidelines for the conduct of clinical trials in pain patients usually recommend long study periods, in addition to the several weeks of medication adjustments that are often necessary to reach steady-state conditions, making clinical trials in this indication a costly and time-consuming process [8]. Thus, prior evaluation of the efficacy of new drugs or alternative methods for pain relief in human surrogate models in healthy volunteers may serve as an initial proof of concept, while they may also help to improve the study design and to define relevant efficacy parameters in subsequent clinical trials.

1

Several forms of nociceptive activation have been used in experimental models of sensitization in humans, among which electrical stimulation and chemical irritation using a variety of substances (e.g. capsaicin, mustard oil) have frequently been used [4,9-12,15,17]. The assessment of central sensitization effects produced by these models is usually carried out using psychophysical measures, based on a subjective evaluation performed by the volunteers [5]. However, the nociceptive withdrawal reflex (NWR) appears as an excellent alternative for the assessment of central sensitization within this context. It is an objective, electrophysiological measure of spinal nociception, highly correlated to pain in healthy volunteers and in several pain syndromes in patients [3,16]. Moreover, derived measures such as the reflex receptive fields (RRF) can provide additional information about functional characteristics of the NWR under different conditions [1].

1.1 AIMS OF THE PH.D. PROJECT

The aims of this Ph.D. project were: 1) to explore different models of central sensitization in humans and 2) to assess these models objectively using the NWR and the RRF.

Specifically, the research questions addressed in this project were:
1. Is it possible to improve the assessment of NWR and RRF in humans?
2. How reliable are the NWR and RRF as objective measures of spinal nociception?
3. What are the parameters that influence the induction and establishment of human surrogate models of central sensitization?
4. Are the NWR and RRF able to assess the effects of central sensitization models in humans?

These questions are addressed throughout eight peer-reviewed articles, divided in four main studies (from now on referred to as Study I to IV), and four supplementary papers (referred to as SP I to IV).

The four main studies are:

Study I
Biurrun Manresa JA, Jensen MB, Andersen OK (2011) *Introducing the reflex probability maps in the quantification of nociceptive withdrawal reflex receptive fields in humans.* J Electromyogr Kines 21:67-76. DOI:10.1016/j.jelekin.2010.09.003

Study II
Biurrun Manresa JA, Neziri AY, Curatolo M, Arendt-Nielsen L, Andersen OK (2010) *Test-retest reliability of the nociceptive withdrawal reflex and electrical pain thresholds after single and repeated stimulation in patients with chronic low back pain.* Eur J Appl Physiol 111:83-92. DOI: 10.1007/s00421-010-1634-0

Study III
Biurrun Manresa JA, Mørch CD, Andersen OK (2010) *Long-term facilitation of nociceptive withdrawal reflexes following low-frequency conditioning electrical stimulation: A new model for central sensitization in humans.* Eur J Pain 14:822-831. DOI: 10.1016/j.ejpain.2009.12.008

Study IV
Biurrun Manresa JA, Finnerup NB, Johannesen IL, Biering-Sørensen F, Jensen TS, Arendt-Nielsen L, Andersen OK (2011) *Expansion of nociceptive withdrawal reflex receptive fields in complete spinal cord injured patients and healthy volunteers during capsaicin-induced central sensitization.* Submitted to J Neurosci.

The four supplementary papers are:

SP I
Biurrun Manresa JA, Hansen J, Andersen OK (2010) *Development of a data acquisition and analysis system for nociceptive withdrawal reflex and reflex receptive fields in humans.* Proc 32nd Annual International Conference of the IEEE Engineering in Medicine and Biology Society IEEE EMBS 2010. Buenos Aires, Argentina, August 31 - September 4, 2010. ©IEEE, pp. 6619-6624.

SP II
Biurrun Manresa JA, Mørch CD, Andersen OK (2010) *Teager-Kaiser energy operator improves the detection and quantification of nociceptive withdrawal reflexes from surface electromyography.* Proc 18th European Signal Processing Conference EUSIPCO 2010. Aalborg, Denmark, 23-27 August. ©EURASIP ISBN 2076-1465, pp. 910-913.

SP III
Neziri AY, Haesler S, Petersen-Felix S, Müller M, Arendt-Nielsen L, **Biurrun Manresa JA**, Andersen OK, Curatolo M (2010) *Generalized expansion of nociceptive reflex receptive fields in chronic pain patients.* Pain 151:798-805. DOI: 10.1016/j.pain.2010.09.017

SP IV
Van Den Broeke EN, Van Rijn CM, **Biurrun Manresa JA**, Andersen OK, Arendt-Nielsen L, Wilder-Smith OHG (2010) *Neurophysiological correlates of nociceptive heterosynaptic long-term potentiation in humans.* J Neurophysiol 103:2107-2113. DOI: 10.1152/jn.00979.2009

1.2 DISSERTATION OVERVIEW

This thesis describes the methodology for the induction and establishment of human models of central sensitization and the assessments of the effects of these models using the NWR and RRF, as reported in the studies mentioned before. The link between these studies can be seen in fig. 1.1.

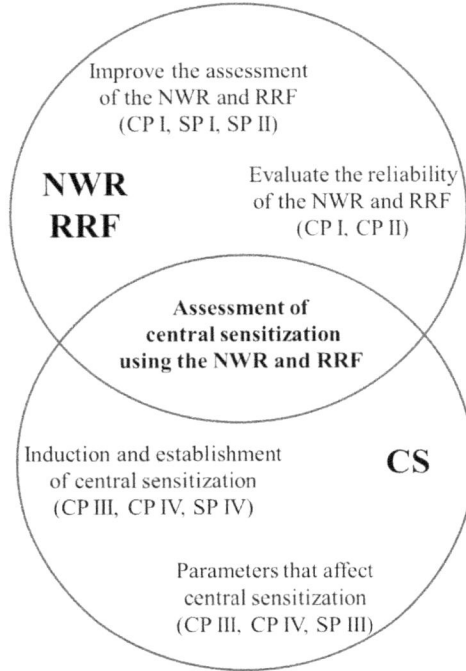

Fig. 1.1. Dissertation overview.

The assessment of the NWR and RRF in humans (question no. 1) is addressed in Study I, SP I and SP II. The reliability of these methods (question no. 2) is established in Study I and Study II. The parameters that influence the induction and establishment of human surrogate models of central sensitization (question no. 3) are investigated in Study III, Study IV and SP IV. Finally, the possibility of using the NWR and RRF as assessment methods for different models of central sensitization (question no. 4) is addressed in Study III, Study IV and SP III.

REFERENCES

[1]   Andersen OK. Studies of the organization of the human nociceptive withdrawal reflex. Focus on sensory convergence and stimulation site dependency. Acta Physiol 2007;189:1-35.

[2]   Cook AJ, Woolf CJ, Wall PD, McMahon SB. Dynamic receptive field plasticity in rat spinal cord dorsal horn following C-primary afferent input. Nature 1987;325:151-153.

[3]   García-Larrea L, Mauguière F. Electrophysiological assessment of nociception in normals and patients: the use of nociceptive reflexes. Electroencephalogr Clin Neurophysiol Suppl 1990;41:102-118.

[4]   Grönroos M, Pertovaara A. Capsaicin-induced central facilitation of a nociceptive flexion reflex in humans. Neurosci Lett 1993;159:215-218.

[5]   Hansen N, Klein T, Magerl W, Treede R-. Psychophysical evidence for long-term potentiation of C-fiber and Aδ-fiber pathways in humans by analysis of pain descriptors. J Neurophysiol 2007;97:2559-2563.

[6]   Herrero JF, Laird JMA, Lopez-Garcia JA. Wind-up of spinal cord neurones and pain sensation: Much ado about something? Prog Neurobiol 2000;61:169-203.

[7]   Ji RR, Kohno T, Moore KA, Woolf CJ. Central sensitization and LTP: do pain and memory share similar mechanisms? Trends Neurosci 2003;26:696-705.

[8]   Klein T, Magerl W, Hanschmann A, Althaus M, Treede R-. Antihyperalgesic and analgesic properties of the N-methyl-d-aspartate (NMDA) receptor antagonist neramexane in a human surrogate model of neurogenic hyperalgesia. Eur J Pain 2008;12:17-29.

[9]   Klein T, Magerl W, Hopf HC, Sandkühler J, Treede R-. Perceptual Correlates of Nociceptive Long-Term Potentiation and Long-Term Depression in Humans. J Neurosci 2004;24:964-971.

[10] Koltzenburg M, Lundberg LER, Torebjork HE. Dynamic and static components of mechanical hyperalgesia in human hairy skin. Pain 1992;51:207-218.

[11] Koppert W, Dern SK, Sittl R, Albrecht S, Schu☐ttler J, Schmelz M. A new model of electrically evoked pain and hyperalgesia in human skin: The effects of intravenous alfentanil, S(+)-ketamine, and lidocaine. Anesthesiol 2001;95:395-402.

[12] LaMotte RH, Shain CN, Simone DA, Tsai E-P. Neurogenic hyperalgesia: Psychophysical studies of underlying mechanisms. J Neurophysiol 1991;66:190-211.

[13] Latremoliere A, Woolf CJ. Central Sensitization: A Generator of Pain Hypersensitivity by Central Neural Plasticity. J Pain 2009;10:895-926.

[14] Loeser JD, Treede R-. The Kyoto protocol of IASP Basic Pain Terminology. Pain 2008;137:473-477.

[15] Magerl W, Wilk SH, Treede R-. Secondary hyperalgesia and perceptual wind-up following intradermal injection of capsaicin in humans. Pain 1998;74:257-268.

[16] Sandrini G, Serrao M, Rossi P, Romaniello A, Cruccu G, Willer JC. The lower limb flexion reflex in humans. Prog Neurobiol 2005;77:353-395.

[17] Torebjork HE, Lundberg LER, LaMotte RH. Central changes in processing of mechanoreceptive input in capsaicin-induced secondary hyperalgesia in humans. J Physiol 1992;448:765-780.

[18] Woolf CJ. Evidence for a central component of post-injury pain hypersensitivity. Nature 1983;306:686-688.

[19] Woolf CJ, Salter MW. Neuronal Plasticity: Increasing the Gain in Pain. Science 2000;288:1765-1768.

[20] Woolf CJ, Wall PD. Relative effectiveness of C primary afferent fibers of different origins in evoking a prolonged facilitation of the flexor reflex in the rat. J Neurosci 1986;6:1433-1442.

Chapter   2.

Objective assessment of central sensitization

The assessment of the effects of central sensitization in experiments involving human participants is a challenging task. A widely used option is to examine specific somatosensory changes in pain perception after conditioning stimulation, assessing the state of the entire nociceptive system using methods based on subjective responses [9,40,46]. The NWR, on the other hand, is an objective electrophysiological measure commonly used to assess spinal processing of nociception in animal [47] and human experiments, where it has been extensively applied in studies involving healthy volunteers as wells as in the research of chronic pain conditions and other painful disorders [1,70].

2.1 THE NOCICEPTIVE WITHDRAWAL REFLEX

The NWR is a typical defense reaction observed in almost all living species, with the purpose of withdrawing the extremities from potential damaging stimuli. Sherrington first described this response in animals at the beginning of the 20th century, running a series of experiments where noxious electrical stimulation of the limbs caused a stereotyped flexion of the stimulated limb to withdraw it from the stimulus, associated with an extension of the contralateral limb to preserve balance [75]. He named this pattern *flexion reflex*, although later research showed that an extension reflex could also be elicited depending on the site where the stimulus was applied [39], thus expanding the concept to the more general term *withdrawal reflex*.

2.1.1   Stimulation and recording of the NWR

A NWR can be elicited by natural and artificial stimuli. Examples of natural stimuli are heat and mechanical punctuate stimuli, which activate specific pain receptors in the skin [57,73,89]. Although the NWR elicited by these stimuli could be easily associated with responses in natural conditions (e.g. stepping on a sharp object or touching a hot plate), they present a few methodological disadvantages, such as the impossibility to rely on accurate timing from the onset of the stimulus until the response is measured or potential tissue damage after repeated stimulation. On the other hand, electrical stimulation is the most widely used artificial method for eliciting the NWR, since it is easier to control and deliver [85]. Moreover, this kind of stimulus bypasses the skin receptor and generates a synchronous action potential directly in the sensory nerve, resulting in highly reproducible reflexes in comparison with other methods, such as radiant heat [57].

The afferent barrage eliciting the NWR depends on the anatomical structures being stimulated: stimulation of a nerve trunk / bundle will likely produce an afferent barrage consisting of cutaneous component from the stimulated skin, plus components from afferents innervating distal skin, proprioceptors, muscles, joint capsules and deep structures [53]. Localized stimulation of the skin likely depolarizes thin myelinated and unmyelinated fibers, although components from other structures cannot be discarded. In both cases, the terms RII and RIII were introduced to characterize reflexes evoked by group II (Aβ) and group III (Aδ) fibers respectively, usually differentiated by the reflex onset latency [41]. In all the studies presented in this thesis, the volunteers described the electrical stimulus as a sharp, pricking, and well localized sensation, most likely reflecting Aδ afferent inflow [36].

In any case, electromyography (EMG) is commonly used to record the NWR response from the muscles [35,39,70,75]. There are two different recording strategies for EMG: invasive, in which a direct measurement of muscle fibre activity is obtained by intra-muscular needle electrodes, and non-invasive, where integrated potentials are acquired by surface electrodes placed on the skin. In humans, surface EMG recording is generally preferred. The most important advantage of surface EMG is that it is not necessary to insert needles into the muscle, avoiding damage the muscle tissue during a contraction and risk of infection. Moreover, the insertion of needles can change the sensory inflow to the spinal cord and therefore affect the spinal control. However, surface EMG has the disadvantage of possible contamination by noise, e.g., ambient and transducer noise, artefacts and unwanted signals from other muscles in close proximity to the muscle fibres of interest, namely myoelectric cross-talk [23]. For more details refer to SP I, which presents a description of a data acquisition and analysis system for NWR in humans.

2.1.2 Detection and quantification of the NWR

Several methods for detection and quantification of the NWR in surface EMG recordings have been introduced, e.g., integrated and mean EMG amplitude [18], area under the curve [29], maximal peak to peak amplitude [45], and root-mean-square (RMS) amplitude [6], among others. In particular, it is worth mentioning the efforts of Rhudy, France and colleagues towards a standard definition of the NWR threshold using the best possible scoring criteria for NWR detection [34,66]. Nevertheless, the performance of all detection and quantification methods is negatively affected when the surface EMG signal is contaminated with noise. Most of the methods developed to overcome this difficulty are complex and computationally intense, and often *a priori* knowledge of the properties of the surface EMG signals is required [49].

In SP II, a fast and simple method to improve the characterization of the NWR was proposed. It consisted on pre-processing the surface EMG signals with the Teager-Kaiser energy operator (TKEO) prior to the detection and quantification stage. The algorithm is based on a nonlinear operator that tracks the energy of the system that produces a signal instead of the signal's energy itself [43]. A subset of NWR data from 300 healthy volunteers, recorded from tibialis anterior (TA) and soleus (SOL) muscles, was used to evaluate several methods for reflex detection and quantification, compared with and without TKEO pre-processing. Receiver operating characteristic (ROC) analysis was carried out to determine the performance of each method while detecting the NWR, by comparison to NWR detection performed by an expert. The results showed a significant improvement on NWR detection when the TKEO operator was used to pre-process the EMG signals.

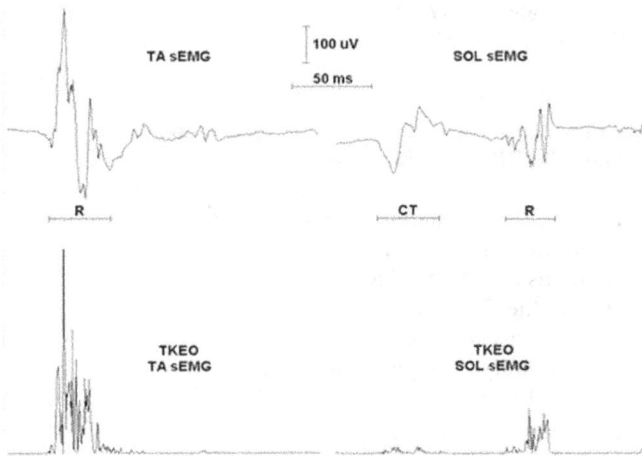

Fig. 2.1 Surface EMG (sEMG) signals before and after pre-processing (**R**: reflex, **CT**: cross-talk).

ROC analysis showed a good performance of all methods in the detection of the NWR, in agreement with previous studies [66]. Methods involving peak values performed best, with areas under the ROC curve greater than 0.92. There is a noticeable difference between performances in TA recordings compared to SOL recordings: NWR detection in TA is in average 5% better than in SOL. This is to be expected because SOL signals are more affected by cross-talk and noise than TA signals, due to the fact that the most common withdrawal pattern is dorsiflexion of the ankle, which mostly involves TA activity [1]. Nevertheless, this difference disappears when TKEO pre-processing is applied (with improvements up to 12% in some cases), and all methods accomplish areas under the ROC curve greater than 0.95, therefore becoming reliable for NWR detection task.

Since there is not an objective pattern to measure the accuracy of quantification for any method, a comparison cannot be established. Previous work using both simulated surface EMG models and experimental data showed that the frequency content of the signal recorded alone cannot give any indication on cross-talk, and as a consequence, cross-talk reduction cannot be achieved by temporal high-pass filtering only [31,42]. In the light of these results, it could be argued that if the detection improves after pre-processing the recordings with the TKEO (taking into account both amplitude and frequency content), it must be due to a reduction in the effect of noise and cross-talk over the signals, that is, an enhancement in the signal-to-noise ratio (as can be seen on fig. 2.1). Thus, if the signal-to-noise ratio improves, then the quantification process should be more accurate, leading to a better characterization of the NWR.

2.2 THE REFLEX RECEPTIVE FIELD

Studies in both animals [20,37,73,86] and humans [3,77] have demonstrated a modular organization in the spinal control of the nociceptive withdrawal reflexes, meaning that each muscle or group of synergistic muscles has a well-defined and unique cutaneous reflex receptive field (RRF). Noxious stimulation of the skin within the RRF may cause a reflex response involving the related muscles, whereas stimulation outside the RRF may have no effect or may even inhibit activity in the same muscles [76,86]. The RRF is hence defined as the skin area from which a reflex can be evoked, which generally adheres to the biomechanical function of the related group of muscles ensuring adequate withdrawal [20,73,86].

Several studies assessed the RRF of lower limb muscles in humans using electrical stimulation, the first one published more than a decade ago [3]. From that starting point, many aspects of the RRF have been studied: the modular organization of excitatory and inhibitory receptive fields [76], the sensory convergence of painful and non-painful inputs [4] and the modulation of RRF by several parameters, such as ongoing motor programme and stimulation site, phase and frequency, among others [5,6,79,81]. In time, this led to the development of a method for quantification of the RRF based on bidimensional interpolation and extrapolation of EMG amplitudes (fig. 2.2). A set of derived features describing the size and location of the RRF can be derived for each muscle [59], from which the RRF area appears to be the most representative parameter [58]. A thorough description of a data acquisition and analysis system for RRF in humans is also presented in SP I.

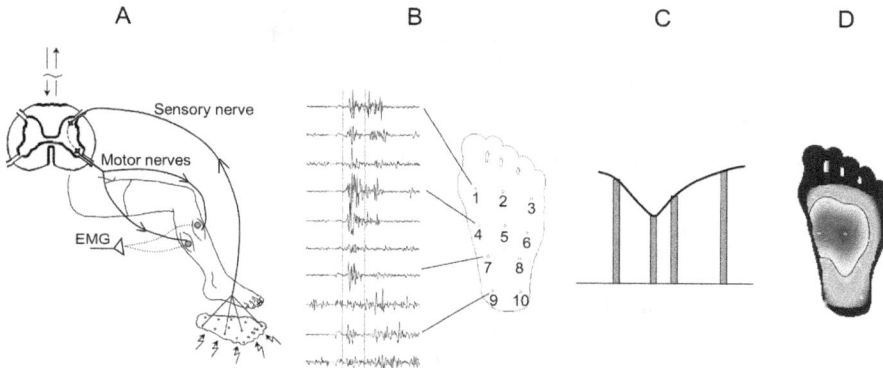

Fig. 2.2. General method for obtaining RRF. **A** NWR responses are evoked by distributed electrical stimulation on the sole of the foot using surface electrodes at distinct locations. The reflex responses are recorded by surface EMG. This could be at any muscle biomechanically involved in the NWR response. **B** Stimuli are delivered at all sites in randomised sequence, and the EMG signals are averaged for every stimulation site. The reflex size is quantified in the 60–180 ms time interval (indicated by the vertical lines). **C** The NWR size detected at each electrodes is interpolated and extrapolated. **D** The two-dimensional interpolation map is then superimposed onto a map of the foot for depicting the NWR sensitivity in a particular muscle (modified from Neziri et al., 2009).

Since it is a recently developed technique, variations in the stimulation parameters needed to be explored in order to investigate their influence on several variables that can affect RRF assessment, e.g., stimulation intensity, subjective pain perception, stimulus repetition and electrode location, among others. An optimal combination of these parameters could provide a more reliable assessment, reducing the effect of factors that typically increase the variability of the measurements, such as habituation and changes in vigilance. In addition, only reflex size and, in a lesser degree, onset latencies and joint angles have been used as quantification variables. However, these are not the only measurable factors in NWR analysis; as an example, recent studies have demonstrated that not only the size, but also probability of reflex occurrence could be modulated after conditioning electrical stimulation [28,74]. The advantage of an approach based on probabilities would be that they can be readily obtained and quantified, and the outcome measurements are intrinsically normalized across subjects, resulting in a more general applicability. Thus, a quantification method based on the probability of occurrence, i.e. the RRF probability maps, could provide additional insight into the processes underlying the NWR at a spinal level.

In Study I, repeated electrical stimulation was applied to elicit the NWR in healthy volunteers in order to determine the best parameters for optimal RRF quantification in humans. During two different sessions, fixed (FSI) and adjusted (ASI) stimulation intensities were applied on non-uniformly distributed sites on the foot sole, and pain intensity ratings along with EMG responses were recorded. RRF sensitivity and probability maps were derived using two-dimensional interpolation, and RRF areas were calculated for these maps. The FSI paradigm kept the stimulation intensities constant, but the pain ratings dropped significantly after ten repetitions (fig. 2.3). In contrast, ASI maintained the pain ratings stable, but the stimulation intensities increased significantly after five and ten repetitions (fig. 2.3). However, none of the paradigms altered the RRF areas in a significant way.

2.2.1   Influence of stimulation paradigm in RRF assessment

The RRF reflects the reflex responsiveness as a function of the stimulation site, and it is often interpreted as the sensitivity of the spinal reflex pathways. This assumption implies that there is no spatial dependency in the sensitivity related to the stimulation site; thus, factors like variations in skin thickness and nerve innervation density must be considered carefully in order to select appropriate stimulation intensities [1]. One possible way to accomplish this is to titrate the stimulus intensity to the pain threshold at every electrode site, which can be done in different ways. The FSI paradigm resembles procedures previously used in several studies [6,59,76,80]. It assumes linearity of the stimulus-response functions for the various stimulation sites, i.e., that multiplication of the intensity corresponding to the pain threshold by a fixed factor entails uniform pain intensity. This is an indirect method for accomplishing equal afferent input, as it assumes equal sensitivity in the ascending sensory pathways and in the reflex encoding pathways and ignores peripheral stimulus-response differences between sites [1].

The ASI paradigm is not based on the linearity assumption; instead, the pain threshold is determined at a single site (e.g. arch of the foot), a multiplication factor is applied and only afterwards the rest of the sites are assessed until homogeneous pain intensity level is obtained across sites. Additionally, the intensities were reassessed after five and ten repetitions were completed at each site, in order to counterbalance central changes that can provoke diminished reflex size and lower pain intensity ratings (e.g. changes in descending activity, habituation). As a result, subjective pain ratings showed a strong relationship with the stimulation paradigm: using the FSI paradigm, the ratings dropped significantly with time, whereas in the ASI paradigm the stimulation intensities had to be steadily increased in order to maintain the pain intensity ratings at a constant level.

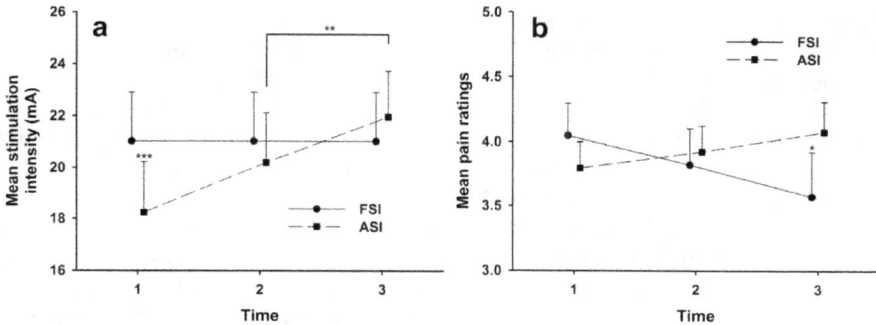

Fig. 2.3. a Mean stimulation intensities across sites as a function of time. Intensities in ASI session at time 1 were significantly lower than intensities at any other session – time combination (*** $p < 0.001$). Intensities in ASI session at time 2 were significantly lower than intensities at time 3 (** $p < 0.01$). **b** Mean pain ratings across sites as a function of time. Pain ratings in FSI session at time 3 were significantly lower than pain ratings at any other any other session – time combination (* $p < 0.05$). Mean + SEM values across 15 volunteers are shown.

FSI throughout the experiment causes decreasing pain intensity sensation with time, probably due to habituation of subjective pain perception to repetitive stimulation [55,84]. This becomes an issue when the subjective pain ratings are used to determine the initial reflex stimulation parameters or when they become the quantifiable outcome variable in human pain models. In the first case, if the pain threshold is used as a reference value, stimulation paradigms that were initially painful might become non-painful within a variable interval of time. In the second case, additional experimental considerations (e.g. supplementary control conditions) must be taken so habituation does not mask the underlying phenomenon under investigation [44,68].

Interestingly, variations due to stimulation paradigms were not observed for RRF measurements. The results in Study I showed the RRF areas were not significantly affected by the stimulation paradigms, and remained stable over time during the course of the experiment. Similar results were already reported in previous reflex studies, where it was discovered that a proper selection of the stimulation parameters, e.g. random inter-stimulus intervals, stimulation of different sites and varying stimulation intensities, can prevent reflexes from habituating [8,27] or can even dishabituate them if habituation already occurred [26,38].

2.2.2   Influence of stimulation sites in RRF assessment

Site dependency of the EMG and kinematic responses of the NWR in humans have been reported for stimulation in sitting position [3], during gait [78], for repetitive stimulation during sitting and standing [6], and as a consequence of pathological conditions [71], among others. In general terms, ankle flexor muscles (primarily tibialis anterior) are activated after stimulation of the medial and distal regions of the sole of the foot, while ankle extensor muscles (mainly soleus and gastrocnemius medialis) are activated after stimulation of the proximal region of the sole of the foot [1]. However, these studies did not investigate in detail the effects of the spatial resolution of stimulated area, and consequently a fixed number of locations (ranging from three to sixteen) was chosen and non-uniformly distributed across the sole of the foot [2,30,77,81].

In Study I, sixteen electrodes were placed in such a way that they covered the entire sole of the foot. In addition to RRF maps, a cluster analysis was performed in order to group the stimulation sites according to two factors depicting similarity: size and probability of occurrence of the reflex. The results consistently remarked a higher sensitivity in the medial region that was singled out in all groupings for both factors. Studies in animals point out that there is no evidence for differences in nociceptor density in the sole of the foot [17,48], so these differences might be primarily due to variations in skin thickness, since below a certain depth primarily thick myelinated fibres are activated and therefore it is difficult to obtain the same amount of thin fiber activation at the heel / central pads [56]. Another outcome of this analysis was the fact that the proximal region entails a significant redundancy regarding information about size or probability: sites at the heel area were always grouped together. There were differences in the groups according to the criterion that was used, reinforcing the idea that size and probability, although still correlated, might convey different information (i.e. they are complementary rather than mutually exclusive measures). Although these results might help in the selection of the number and location of electrodes (e.g., suggesting a higher density of stimulation electrodes in the arc and a smaller density at the heel or variations in innervation depth) in the final decision there are other important factors to weigh, among others the expected or required motor response, the relative level of discomfort and the total duration of the experiment.

2.2.3   Influence of temporal summation in RRF assessment

When comparing single vs. repeated stimulation, it could be noted that larger RRF sensitivity areas were elicited by the 2nd stimulus compared to the 1st stimulus in both paradigms. This observation supports previous findings of temporal summation, indicating graded sensitivity of the NWR [7,8] and the RRF [1,6]. Another factor to be considered is the state of vigilance or awareness [50], since the 1st stimulus acts as a warning for the 2nd one; several studies have shown that this anticipation of strong pain [16,24,88] or the introduction of a warning signal [16,33] induce facilitation of the NWR. RRF probability areas, however, were not significantly affected by the number of stimuli. This does not necessarily mean that the probability of occurrence of the NWR is not affected by temporal summation or anticipation; instead, a likely explanation for these results can be found in the particular choice of threshold for the RRF probability maps or the fact that the frequency of occurrence of the NWR is affected by temporal summation and other central mechanisms on a lesser degree than reflex size (which can be also observed in Study III).

2.3 RELIABILITY OF THE NWR AND RRF

Reliability can be defined as the consistency of measurements on a test [69]. It could be considered as the amount of measurement error that has been deemed acceptable for the effective practical use of a measurement tool. Reliability is essential if a pain test is used for detecting differences between healthy and diseased patients, to follow-up the progression of a given disease in patients, and to investigate the effect of pharmacological interventions, among others. As such, reliability has to be analyzed prior to any other experimental hypothesis, since their validity could be questioned if such test is not adequately consistent in whatever value it indicates from repeated measurements [10].

2.3.1   Methodological aspects of reliability assessment

Two types of reliability can be derived: within-session reliability, also called *internal consistency,* and between-session reliability, also referred to as *stability over time* [13]. The former assesses the reliability of measures that are applied repeatedly during the course of a single session usually within the same day (e.g. before-after experimental designs). The latter evaluates the reliability of measures when repeated experimental sessions are carried out in different days. Both types can be assessed using several methods described below:
   - *Intraclass correlation coefficient (ICC)*: it measures the relative homogeneity within sessions in relation to the total observed variation between sessions. ICC values above 0.75 are indicative of good reliability [62].
   - *Coefficient of variation (CV)*: it represents the standard error of measurement expressed as a percentage of the volunteer's average threshold. The CV can be interpreted as the percentage of deviation from the average threshold below which 68% of the differences between sessions may be expected to lie [10].
   - *Bland-Altman agreement analysis*: it is based on the analysis of the average vs. the difference of the thresholds between two given sessions, from which the so called *limits of agreement (LA)* can be derived, as the average difference ± 1.96 times the standard deviation of the differences. The LA delimit the range within which 95% of the differences between thresholds in two single sessions may be expected to lie. In close relation to this definition, the *coefficient of repeatability (CR)* is defined as the value below which 95% of the *absolute* differences between thresholds in two single sessions may be expected to lie [15]. A graphical interpretation of some of these reliability measures can be seen in fig. 2.4.

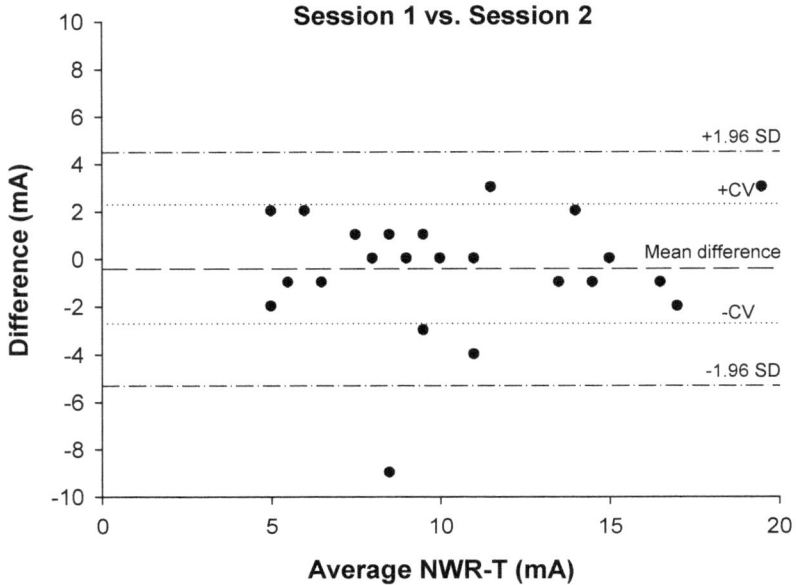

Fig. 2.4. Graphical illustration of reliability measures. The Bland-Altman plot depicts the relationship between the difference and the average of a particular measure (in this case, the nociceptive withdrawal reflex threshold, NWR-T) assessed at two different time points (in this case, two sessions that were one week apart in time). The dashed line represents the mean difference (that should be close to zero if there is no bias between sessions); the dotted line represents the coefficient of variability (CV) and the dashed-dotted line represents the limits of agreement (LA).

The assessment of test-retest reliability and the comparisons of results from different studies should be done cautiously, depending on the type of parameter used to measure reliability [10]. The ICC has an advantage over other correlation methods (such as Pearson's correlation coefficient), because it can be used when more than one retest is performed. However, these methods depend on the sample heterogeneity [14], and thus are considered measures of relative reliability, since the more homogeneous a population is, the lower measurement error is needed in order to detect differences between individuals in a population [10]. In contrast, measures of absolute reliability (such as standard error of measurement, CV and LA) are not affected by the range of measurements in use. The standard error of measurement and the LA are reported in the same dimension (i.e. units) of the test, whereas the CV is a dimensionless statistic and thus it is useful to compare the reliability among studies using different methodologies [32].

2.3.2   Reliability of the NWR

Although several parameters can be employed to describe the NWR (e.g. amplitude, latency, RMS), one of the most frequently used ones is the NWR threshold (NWR-T), defined as the smallest stimulation intensity that elicits a reflex. Moreover, the NWR-T is usually assessed in connection to the electrical pain threshold (EP-T), i.e., the smallest stimulation intensity that elicits a painful sensation. Previous studies addressed the reliability between the NWR-T and EP-T mainly in populations of healthy volunteers. Dincklage et al. [27] reported that the variability between test and retest of the NWR-T after single stimulation, measured as the standard deviation of the differences between measurements, was approximately 4.4 mA when the sessions were approximately 16 weeks apart. Micalos et al. [54] reported that the reliability analysis of the NWR-T after single stimulation showed in average a CV of 16.9% and an ICC of 0.82, whereas for EP-T, also after single stimulation, the values were in average a CV of 16.1% and an ICC of 0.88, when the sessions were separated approximately by 4 days. Similar conclusions were also reached by Lund et al. [51] in a study involving healthy volunteers and pain patients; however only sensory and pain thresholds to electrocutaneous stimulation were tested, and a custom-designed device with an ordinal scale was used, thus making it difficult to compare these results against similar studies. Both studies concluded that the NWR-T and EP-T are reliable measurements in healthy volunteers, and therefore can be applied as tools in experimental pain studies.

In connection with the assessment of central sensitization, it was also necessary to assess population groups that display pain hypersensitivity, in order to confirm that these methods are still reliable in such situations. In Study II, the aims were to determine the test-retest reliability of the NWR-T and EP-T after single and repeated (temporal summation) electrical stimulation in a group of patients with chronic low back pain, and to investigate the association between the NWR-T and the EP-T. Three identical sessions were carried out, separated in average by one week, where the NWR-T and EP-T after single and repeated stimulation were assessed. The results showed that the NWR-T was significantly higher than the EP-T and that the thresholds obtained after single stimulation were significantly higher than those obtained after repeated stimulation, but no significant differences (bias) were found between sessions. Both NWR-T and EP-T presented good to excellent test-retest reliability, as can be seen in table 2.1. After repeated stimulation, the reliability values were similar for NWR-T and EP-T and generally lower when compared to the results obtained after single stimulation. Threshold reliability was highest when the assessment was done between the second and third sessions and lowest between the first and the last sessions. Finally, the association between the NWR-T and EP-T was better for repeated stimulation than for single stimulation.

Human Models of Central Sensitization

Table 2.1. Detailed analysis for ICC and CV (* $p < 0.05$; ** $p < 0.01$; *** $p < 0.001$ in F test for ICC with hypothesized value of 0.5). For Bland-Altman analysis results, refer to Study II in the appendix.

| | Intraclass correlation (ICC) | | | | | |
| | NWR threshold (NWR-T) | | | Electrical pain threshold (EP-T) | | |
	Sessions 1-2	Sessions 2-3	Sessions 1-3	Sessions 1-2	Sessions 2-3	Sessions 1-3
Single stimulation	0.82 **	0.85 ***	0.71	0.91 ***	0.94 ***	0.84 **
Repeated stimulation	0.80 **	0.84 **	0.62	0.81 **	0.85 **	0.68
	Coefficient of variation (CV)					
	NWR threshold (NWR-T)			Electrical pain threshold (EP-T)		
	Sessions 1-2	Sessions 2-3	Sessions 1-3	Sessions 1-2	Sessions 2-3	Sessions 1-3
Single stimulation	16.8%	14.4%	22.0%	11.4%	9.4%	15.2%
Repeated stimulation	14.8%	13.4%	22.4%	12.7%	12.5%	18.8%

The results in Study II rendered similar reliability values in comparison with studies involving healthy volunteers. In particular, the EP-T appears to have slightly better reliability than the reflex threshold after single stimulation. A possible explanation lies in the fact that the nociceptive input that ultimately elicits the NWR is largely processed in the spinal cord subjected to descending modulation from supraspinal structures [1], whereas report of a painful sensation is subjected to further processing in the brain, that integrates this nociceptive input with additional cognitive and perceptual information [63,64]. Thus, several other variables play an important role in pain perception, and some of them (e.g. habituation to electrical stimulation, attention, memory of the ratings of previous stimulations) can affect it in such a way that the overall variability of the pain ratings is decreased, resulting in an increase of the repeatability (e.g. volunteers tend to repeat the same scores if many ratings are requested). Interestingly, this effect is not so remarkable for temporal summation, probably due to the fact that repeated stimulation provides a more stable, long-lasting nociceptive input that might allow a more reliable reflex response and a better assessment of the pain sensation.

It should be noted that the differences in reliability among the different tests were in general modest. Therefore, it cannot be ruled out that at least some of these differences were the result of chance. In general, the reliability was good to excellent for all tests. Lastly, and although there is no systematic bias in the average NWR-T and EP-T between sessions, the reliability is best for the last two sessions and worst when the first and the last sessions are used for the assessment, possibly suggesting a learning effect [72] or gradually lower vigilance despite the initial familiarization with the experimental procedures. Thus, it is expected that the estimated reliability of the NWR-T and EP-T will improve with an increasing number of sessions and a smaller interval of time between sessions (for instance, in crossover studies). Finally, special caution should be taken when follow-up reliability studies are planned involving long periods of time between sessions.

The reliability of the NWR-T and EP-T obtained in studies involving healthy volunteers appear to be comparable to those presented in Study II for patients with chronic pain. Moreover, a number of studies have addressed the reliability of other tests that are also used to assess somatosensory function (including cutaneous and deep pain sensitivity), such as the quantitative sensory test, QST [9,52,67]. QST test have been widely used to test for sensory differences in a variety of human pain syndromes, such as low back pain [60], whiplash [83], irritable bowel syndrome [87], endometriosis [11], and other pain states [22]. In a recent review, Chong and Cros [19] presented a meta-analysis of the reproducibility of several QST methods (vibration perception threshold, heat-electrical pain threshold, cold perception threshold, and warm perception threshold) in healthy volunteers as well as in patients suffering from pathological conditions (diabetic patients with or without neuropathy), concluding that these tests appeared to be sufficiently reproducible during short-term studies (intervals ranging from 1 to 8 weeks). In comparison to the values exhibited by these methods, the reliability of the NWR-T and EP-T reported here is similar or even better, therefore making them suitable for clinical use.

2.3.3   Reliability of the RRF

The reliability of the RRF methodology was addressed for the first time in Study I. The results showed that RRF area measurements presented high within-session reliability for all assessment methodologies, ranging from good to excellent depending on the specific stimulation paradigm being used. Moreover, the RRF area estimation error also showed acceptable values when five or more repetitions were used for the estimation. In all cases, the error remained under 10% after five repetitions, and under 5% after ten repetitions (fig. 2.5). In a large study set out to establish normative values for NWR and RRF in a population of 300 healthy volunteers, the standard deviation of the RRF area was found to be 17% [58]. The estimation error in this study is well below that number, and it decreases significantly when an increasing number of repetitions are used. Furthermore, the number of repetitions can be selected in the light of these results, to match a specific requirement of precision in the estimation for a particular purpose.

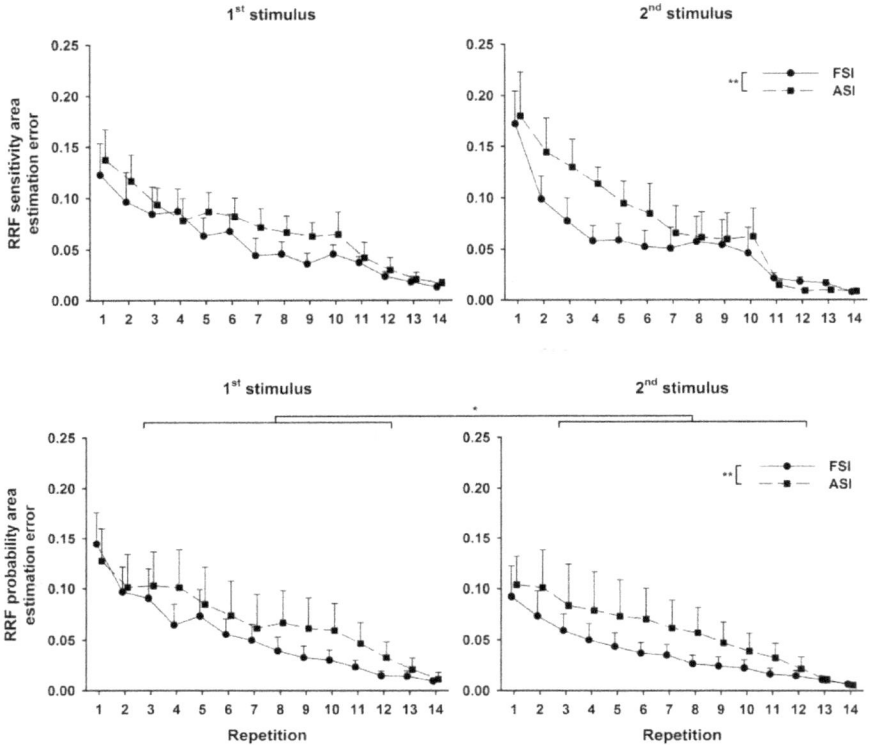

Fig. 2.5. Top RRF sensitivity area estimation error as a function of the number of repetitions. Estimation error for FSI session is significantly smaller than for ASI session (** $p < 0.01$). **Bottom** RRF probability area estimation error as a function of the number of repetitions. Estimation error for FSI session is significantly smaller than for ASI session (** $p < 0.01$). Estimation error for the 2nd stimulus is significantly smaller than for the 1st stimulus (* $p < 0.05$). Mean + SEM values across 15 volunteers are shown.

A remarkable finding in Study I was that the RRF areas obtained after the 2nd stimulus (temporal summation) are more reliable than those obtained after the 1st stimulus, especially when fixed stimulation intensities were used throughout the experiment, most likely due to the fact that the fixed intensities introduce less variability in the RRF assessment, and thus a higher consistency can be achieved in area estimations when focusing on the 2nd stimulus. Since repetitive stimulation does not significantly increase the time required to finish the experiment and does not requires additional considerations either, repeated stimulation appears to be a good way to provide more complete, stable and reliable assessment of RRF parameters without compromising any other aspects of the experiment. Special care must be taken, however, on the selection of the intensities to be used: high stimulation intensities in addition to temporal summation might lead to a decreased range of measurement that could result in saturation, i.e., RRF areas covering the entire sole of the foot [2,6] and potentially induce unacceptable discomfort for the volunteer.

2.4 ASSESSMENT OF CENTRAL SENSITIZATION USING THE NWR AND RRF

Most of the methods used to assess sensitization in humans rely on the volunteer's pain reports after sensory stimulation, which are subjective in nature. A paradigmatic case involves patients suffering from whiplash or fibromyalgia, who present exaggerated pain responses after minimal, undetectable tissue damage following sensory stimulation [21,65]. Nowadays, there is increasing evidence that objective methods, such as the NWR, can detect pain hypersensitivity without the setbacks usually associated to subjective assessments. Indeed, studies involving several patient groups showed that they present lower NWR-T compared to control groups of healthy volunteers [12,25,61,82], which can be interpreted as electrophysiological evidence for hypersensitivity of spinal cord neurons in these patients.

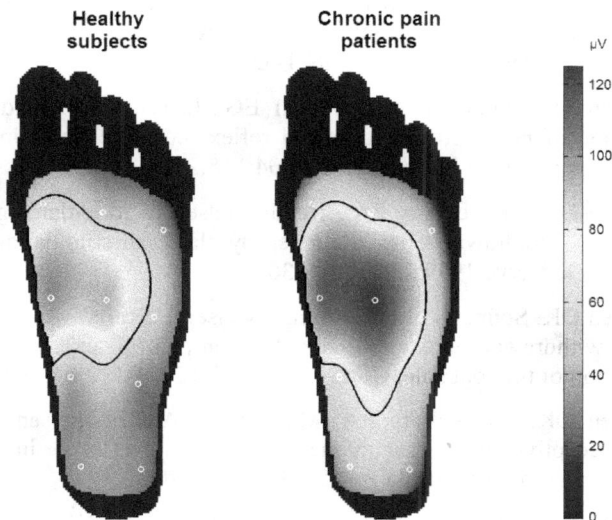

Fig. 2.6. Mean reflex receptive fields (RRF) for healthy volunteers (left) and chronic pelvic pain patients (right). The white dots indicate the stimulation sites. Black line: contour of the RRF area.

In an attempt to corroborate these observations and generalize them to other patient populations, the hypothesis that patients with chronic pelvic pain due to endometriosis display enlarged RRF and lower reflex and pain thresholds compared to pain-free volunteers was tested in SP III. Twenty chronic pain patients and twenty five healthy volunteers participated in the study, in which repeated electrical stimulation was applied on ten sites on the sole of the foot. EMG responses from TA muscle were recorded, from which RRF sensitivity maps were obtained. Additionally, electrical stimulation was applied caudal to the lateral malleolus at the innervation area of the sural nerve, in order to assess NWR-T and EP-T to single and repeated stimulation. The results showed that RRF areas were larger (fig. 2.6) and that NWR-T and EP-T were significantly lower in chronic pain patients compared to healthy volunteers.

These results provide evidence for widespread expansion of spinal neuronal RRF in chronic pain conditions in humans. It is then clear that the NWR and RRF are valuable tools aiming at elucidating the mechanisms that are involved in central sensitization in chronic pain. With that in mind, it is necessary to test if the same conclusion can be achieved using human surrogate models of central sensitization in healthy volunteers, in order to rely on these models for clinical testing, for instance, in the development of new drugs or alternative methods that could potentially alleviate hypersensitivity effects after central sensitization is induced.

REFERENCES

[1]   Andersen OK. Studies of the organization of the human nociceptive withdrawal reflex. Focus on sensory convergence and stimulation site dependency. Acta Physiol 2007;189:1-35.

[2]   Andersen OK, Finnerup NB, Spaich EG, Jensen TS, Arendt-Nielsen L. Expansion of nociceptive withdrawal reflex receptive fields in spinal cord injured humans. Clin Neurophysiol 2004;115:2798-2810.

[3]   Andersen OK, Sonnenborg FA, Arendt-Nielsen L. Modular organization of human leg withdrawal reflexes elicited by electrical stimulation of the foot sole. Muscle Nerve 1999;22:1520-1530.

[4]   Andersen OK, Sonnenborg FA, Arendt-Nielsen L. Reflex receptive fields for human withdrawal reflexes elicited by non-painful and painful electrical stimulation of the foot sole. Clin Neurophysiol 2001;112:641-649.

[5]   Andersen OK, Sonnenborg FA, Matjacic Z, Arendt-Nielsen L. Foot-sole reflex receptive fields for human withdrawal reflexes in symmetrical standing position. Exp Brain Res 2003;152:434-443.

[6]   Andersen OK, Spaich EG, Madeleine P, Arendt-Nielsen L. Gradual enlargement of human withdrawal reflex receptive fields following repetitive painful stimulation. Brain Res 2005;1042:194-204.

[7]   Arendt-Nielsen L, Brennum J, Sindrup S, Bak P. Electrophysiological and psychophysical quantification of temporal summation in the human nociceptive system. Eur J Appl Physiol Occup Physiol 1994;68:266-273.

[8]   Arendt-Nielsen L, Sonnenborg FA, Andersen OK. Facilitation of the withdrawal reflex by repeated transcutaneous electrical stimulation: An experimental study on central integration in humans. Eur J Appl Physiol Occup Physiol 2000;81:165-173.

[9]   Arendt-Nielsen L, Yarnitsky D. Experimental and Clinical Applications of Quantitative Sensory Testing Applied to Skin, Muscles and Viscera. J Pain 2009;10:556-572.

[10] Atkinson G, Nevill AM. Statistical methods for assessing measurement error (reliability) in variables relevant to sports medicine. Sports Med 1998;26:217-238.

[11] Bajaj P, Bajaj P, Madsen H, Arendt-Nielsen L. Endometriosis is associated with central sensitization: A psychophysical controlled study. J Pain 2003;4:372-380.

[12] Banic B, Petersen-Felix S, Andersen OK, Radanov BP, Villiger PM, Arendt-Nielsen L, Curatolo M. Evidence for spinal cord hypersensitivity in chronic pain after whiplash injury and in fibromyalgia. Pain 2004;107:7-15.

[13] Baumgarter TA. Norm-referenced measurement: reliability. In: Safrit MJ, Wood TM. Measurement concepts in physical education and exercise science. Champaigne (IL), USA: Human Kinetics, 1989. pp. 45-72.

[14] Bland JM, Altman DG. A note on the use of the intraclass correlation coefficient in the evaluation of agreement between two methods of measurement. Comput Biol Med 1990;20:337-340.

[15] Bland JM, Altman DG. Measuring agreement in method comparison studies. Stat Methods Med Res 1999;8:135-160.

[16] Boureau F, Luu M, Doubrere JF. Study of experimental pain measures and nociceptive reflex in chronic pain patients and normal subjects. Pain 1991;44:131-138.

[17] Cain DM, Khasabov SG, Simone DA. Response properties of mechanoreceptors and nociceptors in mouse glabrous skin: An in vivc study. J Neurophysiol 2001;85:1561-1574.

[18] Campbell IG, Carstens E, Watkins LR. Comparison of human pain sensation and flexion withdrawal evoked by noxious radiant heat. Pain 1991;45:259-268.

[19] Chong PST, Cros DP. Technology literature review: Quantitative sensory testing. Muscle Nerve 2004;29:734-747.

[20] Clarke RW, Harris J. The organization of motor responses to noxious stimuli. Brain Res Rev 2004;46:163-172.

[21] Curatolo M, Arendt-Nielsen L, Petersen-Felix S. Evidence, mechanisms, and clinical implications of central hypersensitivity in chronic pain after whiplash injury. Clin J Pain 2004;20:469-476.

[22] Curatolo M, Arendt-Nielsen L, Petersen-Felix S. Central Hypersensitivity in Chronic Pain: Mechanisms and Clinical Implications. Phys Med Rehabil Clin North Am 2006;17:287-302.

[23] De Luca CJ, Merletti R. Surface myoelectric signal cross-talk among muscles of the leg. Electroencephalogr Clin Neurophysiol 1988;69:568-575.

[24] Defrin R, Peleg S, Weingarden H, Heruti R, Urca G. Differential effect of supraspinal modulation on the nociceptive withdrawal reflex and pain sensation. Clin Neurophysiol 2007;118:427-437.

[25] Desmeules JA, Cedraschi C, Rapiti E, Baumgartner E, Finckh A, Cohen P, Dayer P, Vischer TL. Neurophysiologic evidence for a central sensitization in patients with fibromyalgia. Arthritis Rheum 2003;48:1420-1429.

[26] Dimitrijević MR, Faganel J, Gregorić M, Nathan PW, Trontelj JK. Habituation: effects of regular and stochastic stimulation. J Neurol Neurosurg Psychiatry 1972;35:234-242.

[27] Dincklage Fv, Hackbarth M, Schneider M, Baars JH, Rehberg B. Introduction of a continual RIII reflex threshold tracking algorithm. Brain Res 2009;1260:24-29.

[28] Don R, Pierelli F, Ranavolo A, Serrao M, Mangone M, Paoloni M, Cacchio A, Sandrini G, Santilli V. Modulation of spinal inhibitory reflex responses to cutaneous nociceptive stimuli during upper limb movement. Eur J Neurosci 2008;28:559-568.

[29] Ellrich J, Treede R-. Convergence of nociceptive and non-nociceptive inputs onto spinal reflex pathways to the tibialis anterior muscle in humans. Acta Physiol Scand 1998;163:391-401.

[30] Emborg J, Spaich EG, Andersen OK. Withdrawal reflexes examined during human gait by ground reaction forces: Site and gait phase dependency. Med Biol Eng Comput 2009;47:29-39.

[31] Farina D, Merletti R, Indino B, Graven-Nielsen T. Surface EMG crosstalk evaluated from experimental recordings and simulated signals. Reflections on crosstalk interpretation, quantification and reduction. Methods Inf Med 2004;43:30-35.

[32] Feltz CJ, Miller GE. An asymptotic test for the equality of coefficients of variation from k populations. Stat Med 1996;15:647-658.

[33] Floeter MK, Gerloff C, Kouri J, Hallett M. Cutaneous withdrawal reflexes of the upper extremity. Muscle Nerve 1998;21:591-598.

[34] France CR, Rhudy JL, McGlone S. Using normalized EMG to define the nociceptive flexion reflex (NFR) threshold: Further evaluation of standardized NFR scoring criteria. Pain 2009;145:211-218.

[35] García-Larrea L, Mauguière F. Electrophysiological assessment of nociception in normals and patients: the use of nociceptive ref.exes. Electroencephalogr Clin Neurophysiol Suppl 1990;41:102-118.

[36] Gardner EP, Martin JH, Jesell TM. The bodily senses. In: Kandell ER, Schwartz JH, Jesell TM. Principles of neural science. New York (USA): McGraw-Hill, 2000. pp. 430-450.

[37] Garwicz M, Jörntell H, Ekerot C-. Cutaneous receptive fields and topography of mossy fibres and climbing fibres projecting to cat cerebellar C3 zone. J Physiol 1998;512:277-293.

[38] Granat MH, Nicol DJ, Baxendale RH, Andrews BJ. Dishabituation of the flexion reflex in spinal cord-injured man and its application in the restoration of gait. Brain Res 1991;559:344-346.

[39] Hagbarth KE. Spinal withdrawal reflexes in the human lower limbs. J Neurol Neurosurg Psychiatry 1960;23:222-227.

[40] Hansen N, Klein T, Magerl W, Treede R-. Psychophysical evidence for long-term potentiation of C-fiber and Aδ-fiber pathways in humans by analysis of pain descriptors. J Neurophysiol 2007;97:2559-2563.

[41] Hugon M. Exteroceptive reflexes to stimulation of sural nerve in man. In: Desmedt JE. New developments in electromyography and clinical neurophysiology. Basel (Switzerland): Kargel, 1973. pp. 713-729.

[42] Jensen MB, Biurrun Manresa JA, Andersen OK. Optimal setup for quantification of nociceptive withdrawal reflex receptive fields in humans. In: Falla D, Farina D. . Aalborg, Denmark: Aalborg University. Department of Health Science and Technology, 2010.

[43] Kaiser JF. On a simple algorithm to calculate the 'energy' of a signal. In: Anonymous Acoustics, Speech, and Signal Processing, 1990. ICASSP-90., 1990 International Conference on, 1990. pp. 381-384.

[44] Klein T, Stahn S, Magerl W, Treede R-. The role of heterosynaptic facilitation in long-term potentiation (LTP) of human pain sensation. Pain 2008;139:507-519.

[45] Koceja DM, Bernacki RH, Kamen G. Methodology for the quantitative assessment of human crossed-spinal reflex pathways. Med Biol Eng Comput 1991;29:603-606.

[46] Lang S, Klein T, Magerl W, Treede R-. Modality-specific sensory changes in humans after the induction of long-term potentiation (LTP) in cutaneous nociceptive pathways. Pain 2007;128:254-263.

[47] Le Bars D, Gozariu M, Cadden SW. Animal models of nociception. Pharmacol Rev 2001;53:597-652.

[48] Leem JW, Willis WD, Chung JM. Cutaneous sensory receptors in the rat foot. J Neurophysiol 1993;69:1684-1699.

[49] Li X, Zhou P, Aruin AS. Teager-Kaiser Energy Operation of Surface EMG Improves Muscle Activity Onset Detection. Ann Biomed Eng 2007;35:1532-1538.

[50] Liebermann DG, Defrin R. Characteristics of the nociceptive withdrawal response elicited under aware and unaware conditions. J Electromyogr Kinesiology 2009;19.

[51] Lund I, Lundeberg T, Kowalski J, Sandberg L, Budh CN, Svensson E. Evaluation of variations in sensory and pain threshold assessments by electrocutaneous stimulation. Physiother Theory Pract 2005;21:81-92.

[52] Magerl W, Krumova EK, Baron R, Tölle T, Treede R-, Maier C. Reference data for quantitative sensory testing (QST): Refined stratification for age and a novel method for statistical comparison of group data. Pain 2010;151:598-605.

[53] Meinck H-, Piesiur-Strehlow B, Koehler W. Some principles of flexor reflex generation in human leg muscles. Electroencephalogr Clin Neurophysiol 1981;52:140-150.

[54] Micalos PS, Drinkwater EJ, Cannon J, Arendt-Nielsen L, Marino FE. Reliability of the nociceptive flexor reflex (RIII) threshold and association with Pain threshold. Eur J Appl Physiol 2009;105:55-62.

[55] Milne RJ, Kay NE, Irwin RJ. Habituation to repeated painful and non-painful cutaneous stimuli: A quantitative psychophysical study. Exp Brain Res 1991;87:438-444.

[56] Mørch CD, Hennings K, Andersen OK. Modeling preferential activation of Aδ-fibers fibers by small-area surface electrodes. In: Anonymous , 2009. pp. No. 763.19/35.

[57] Mørch CD, Andersen OK, Graven-Nielsen T, Arendt-Nielsen L. Nociceptive withdrawal reflexes evoked by uniform-temperature laser heat stimulation of large skin areas in humans. J Neurosci Methods 2007;160:85-92.

[58] Neziri AY, Andersen OK, Petersen-Felix S, Radanov B, Dickenson AH, Scaramozzino P, Arendt-Nielsen L, Curatolo M. The nociceptive withdrawal

reflex: Normative values of thresholds and reflex receptive fields. Eur J Pain 2010;14:134-141.

[59] Neziri AY, Curatolo M, Bergadano A, Petersen-Felix S, Dickenson A, Arendt-Nielsen L, Andersen OK. New method for quantification and statistical analysis of nociceptive reflex receptive fields in humans. J Neurosci Methods 2009;178:24-30.

[60] O'Neill S, Manniche C, Graven-Nielsen T, Arendt-Nielsen L. Generalized deep-tissue hyperalgesia in patients with chronic low-back pain. Eur J Pain 2007;11:415-420.

[61] Perrotta A, Sandrini G, Serrao M, Buscone S, Tassorelli C, Tinazzi M, Zangaglia R, Pacchetti C, Bartolo M, Pierelli F, Martignoni E. Facilitated temporal summation of pain at spinal level in Parkinson's disease. Movement Disorders 2010:n/a-n/a.

[62] Portney LG, Watkins MP. Foundations of clinical research : applications to practice. Upper Saddle River, N.J.: Pearson/Prentice Hall, 2009.

[63] Price DD. Psychological and neural mechanisms of the affective dimension of pain. Science 2000;288:1769-1772.

[64] Price DD. Central neural mechanisms that interrelate sensory and affective dimensions of pain. Mol Interv 2002;2:392-403, 339.

[65] Price DD, Staud R, Robinson ME, Mauderli AP, Cannon R, Vierck CJ. Enhanced temporal summation of second pain and its central modulation in fibromyalgia patients. Pain 2002;99:49-59.

[66] Rhudy JL, France CR. Defining the nociceptive flexion reflex (NFR) threshold in human participants: A comparison of different scoring criteria. Pain 2007;128:244-253.

[67] Rolke R, Magerl W, Campbell KA, Schalber C, Caspari S, Birklein F, Treede R-. Quantitative sensory testing: A comprehensive protocol for clinical trials. Eur J Pain 2006;10:77-88.

[68] Rottmann S, Jung K, Ellrich J. Electrical low-frequency stimulation induces homotopic long-term depression of nociception and pain from hand in man. Clin Neurophysiol 2008;119:1895-1904.

[69] Safrit MJ, Wood TM, editors. Measurement concepts in physical education and exercise science. Champaigne (IL), USA: Human Kinetics, 1989.

[70] Sandrini G, Serrao M, Rossi P, Romaniello A, Cruccu G, Willer JC. The lower limb flexion reflex in humans. Prog Neurobiol 2005;77:353-395.

[71] Schmit BD, Hornby TG, Tysseling-Mattiace VM, Benz EN. Absence of Local Sign Withdrawal in Chronic Human Spinal Cord Injury. J Neurophysiol 2003;90:3232-3241.

[72] Schouenborg J. Learning in sensorimotor circuits. Curr Opin Neurobiol 2004;14:693-697.

[73] Schouenborg J, Kalliomäki J. Functional organization of the nociceptive withdrawal reflexes. Exp Brain Res 1990;83:67-78.

[74] Serrao M, Pierelli F, Don R, Ranavolo A, Cacchio A, Curra A, Sandrini G, Frascarelli M, Santilli V. Kinematic and electromyographic study of the nociceptive withdrawal reflex in the upper limbs during rest and movement. J Neurosci 2006;26:3505-3513.

[75] Sherrington CS. Flexion-reflex of the limb, crossed extension-reflex, and reflex stepping and standing. J Physiol 1910;40:28-121.

[76] Sonnenborg FA, Andersen OK, Arendt-Nielsen L. Modular organization of excitatory and inhibitory reflex receptive fields elicited by electrical stimulation of the foot sole in man. Clin Neurophysiol 2000;111:2160-2169.

[77] Sonnenborg FA, Andersen OK, Arendt-Nielsen L, Treede R-. Withdrawal reflex organisation to electrical stimulation of the dorsal foot in humans. Exp Brain Res 2001;136:303-312.

[78] Spaich EG, Andersen OK, Arendt-Nielsen L. Tibialis Anterior and Soleus Withdrawal Reflexes Elicited by Electrical Stimulation of the Sole of the Foot during Gait. Neuromodulation 2004;7:126-132.

[79] Spaich EG, Arendt-Nielsen L, Andersen OK. Modulation of Lower Limb Withdrawal Reflexes during Gait: A Topographical Study. J Neurophysiol 2004;91:258-266.

[80] Spaich EG, Arendt-Nielsen L, Andersen OK. Repetitive Painful Stimulation Produces an Expansion of Withdrawal Reflex Receptive Fields in Humans. Artif Organs 2005;29:224-228.

[81] Spaich EG, Emborg J, Collet T, Arendt-Nielsen L, Andersen OK. Withdrawal reflex responses evoked by repetitive painful stimulation delivered on the sole of the foot during late stance: Site, phase, and frequency modulation. Exp Brain Res 2009;194:359-368.

[82] Sterling M. Differential development of sensory hypersensitivity and a measure of spinal cord hyperexcitability following whiplash injury. Pain 2010;150:501-506.

[83] Sterling M, Jull G, Vicenzino B, Kenardy J. Sensory hypersensitivity occurs soon after whiplash injury and is associated with poor recovery. Pain 2003;104:509-517.

[84] Thompson RF, Spencer WA. Habituation: A model phenomenon for the study of neuronal substrates of behavior. Psychol Rev 1966;73:16-43.

[85] Tørring J, Pedersen E, Klemar B. Standardisation of the electrical elicitation of the human flexor reflex. J Neurol Neurosurg Psychiatry 1981;44:129-132.

[86] Weng H-, Schouenborg J. Cutaneous inhibitory receptive fields of withdrawal reflexes in the decerebrate spinal rat. J PHYSIOL 1996;493:253-265.

[87] Wilder-Smith CH, Schindler D, Lovblad K, Redmond SM, Nirkko A. Brain functional magnetic resonance imaging of rectal pain and activation of endogenous inhibitory mechanisms in irritable bowel syndrome patient subgroups and healthy controls. Gut 2004;53:1595-1601.

[88] Willer JC, Boureau F, Albe-Fessard D. Supraspinal influences on nociceptive flexion reflex and pain sensation in man. Brain Res 1979;179:61-68.

[89] Willer JC, Boureau F, Berny J. Nociceptive flexion reflexes elicited by noxious laser radiant heat in man. Pain 1979;7:15-20.

Chapter 3.

Human models of central sensitization

Several forms of nociceptive activation can evoke central sensitization, as for example heat or inflammation. However, commonly used human models for sensitization involve topical or intradermal chemical irritation or conditioning electrical stimulation onto the skin [20,36,40,41,70]. Indeed, perceptual correlates of central sensitization have been identified after topical or intradermal administration of capsaicin or repetitive conditioning electrical stimulation [38,44,50].

3.1 CONDITIONING ELECTRICAL STIMULATION MODEL

Focusing on conditioning electrical stimulation models, two different paradigms, high- and low-frequency stimulation (HFS and LFS respectively) are often employed, intending to resemble the firing pattern of primary afferent fibers under different pathophysiological conditions. Several *in vitro* and animal *in vivo* experiments have previously demonstrated that these paradigms are effective in eliciting sensitization in spinal nociception [17,32,58]. Recent studies in humans using both HFS and LFS delivered through a special electrode, designed to target nociceptive afferents using high current densities, were able to show perceptual correlates of central sensitization [38].

Study III reports an attempt to establish a model for central sensitization in humans, in which high- and low-frequency conditioning electrical stimulation were applied to the dorsum of the foot of healthy volunteers. Blood flow scans were acquired and perceptual intensity ratings to mechanical stimuli were assessed in the conditioned area and surroundings. In addition, the NWR was elicited within the same innervation area at graded stimulation intensities, in order to obtain an objective correlate of long-term changes in central nociception. Following LFS, a significant long-lasting facilitation of the NWR was observed for all stimulation intensities used, with an increase of 31% in the reflex RMS amplitudes (fig. 3.1),

31

an increase of 22% in the number of reflexes elicited (fig. 3.2) and a decrease of 2% in the reflex latencies. Coincidentally, the blood flow increased up to 80% in the 10 min after conditioning stimulation (fig. 3.3), differing significantly from HFS and Control sessions. No changes in reflex response were observed after HFS or in the Control session, and no significant difference in the blood flow was observed between these two sessions either.

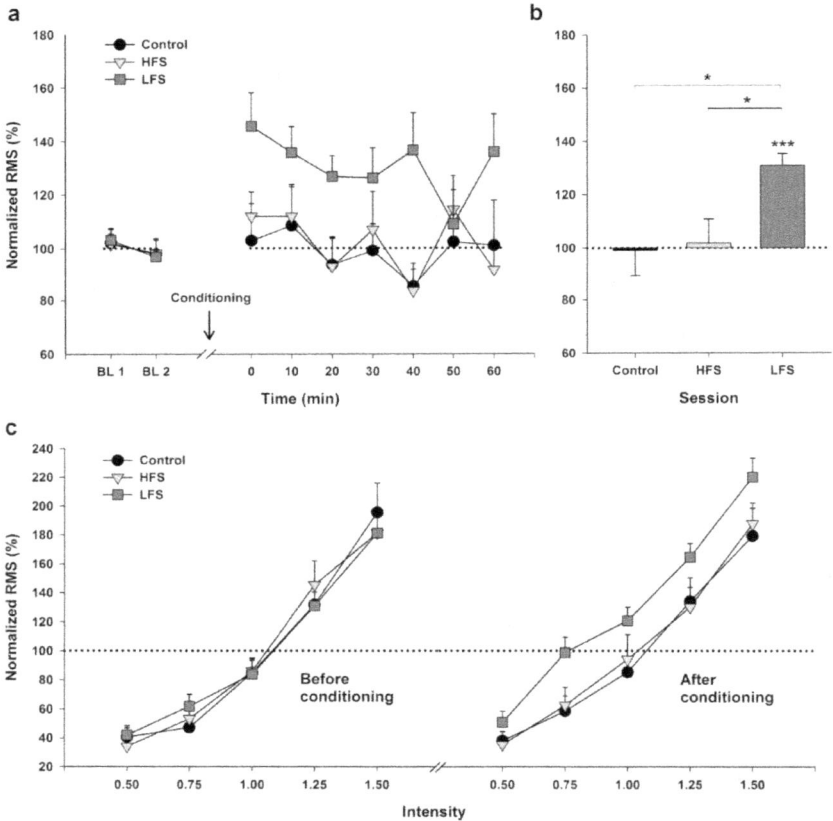

Fig. 3.1. Reflex RMS amplitude. **a** Time course of normalized RMS amplitude before and after conditioning electrical stimulation. **b** Mean values of post-conditioning changes of RMS amplitude across time, in the 0 – 60 min interval; asterisks on top of the bars indicate significant differences (*** $p < 0.001$) on the contrast analysis between pre- and post-conditioning values in each session; asterisks between bars indicate significant differences (* $p < 0.05$) on post-hoc Student-Newman-Keuls tests following RM ANOVA between sessions. **c** Stimulus-response functions for RMS amplitudes, before and after conditioning. Dotted lines indicate mean level of baseline period. Mean ± SEM values across 13 volunteers are shown.

3.1.1   Neural mechanisms of central sensitization

Reflex facilitation with that electrode positioning was likely to be heterotopic, because the conditioning site was different from the test site. However, it is important to note that the conditioning electrode was located within the innervation territory of the superficial peroneal nerve and the reflexes were evoked by compound action potentials of the nerve trunk proximal to the conditioned site, so some fibers could be activated during both conditioning and reflex testing. The design of the conditioning electrode with very small contact surfaces favors activation of nociceptive Aδ and C fibers [33,53], which is assumed to be a prerequisite for induction of central sensitization in this experimental model. The conditioning stimulation intensity (10 times the detection threshold) suggests that Aδ fibers [51] and C fibers [38] would be simultaneously activated. This was corroborated by a significant increase in blood flow after LFS (fig. 3.3), since spreading vasodilatation is correlated to the activation of peptidergic afferents [11]. The electrical test stimulus for evoking NWR, on the other hand, is known to reflect A-fiber activation, as reflected by its onset latencies [1]. Considering the stimulation intensities and stimulation site used in this experiment, probably Aδ fibers were strongly involved. Therefore, the observations in Study III suggest that sensitization is not mediated by an exclusive pathway and also that it is not restricted to a single synapse [45].

Although some homosynaptic phenomena, such as long-term potentiation (LTP), are considered partially responsible for central sensitization [39,59], the essential mechanisms underlying plasticity of somatosensory perception require heterosynaptic interactions of different pathways [9,38,43,54]. Several hypotheses have been proposed to explain such interactions in relation to pain processing in the spinal cord, although they are probably complementary rather than mutually exclusive mechanisms [43]. One of these seems particularly suited to account for the effects observed in Study III; the convergence of A- and C-fiber input onto central nociceptive neurons in the dorsal horn. This convergence has been observed in previous animal [66] and human reflex studies [4], and it is likely to occur in wide-dynamic-range (WDR) neurons located in the deep dorsal horn, which in time are capable of expressing long-term facilitation of synaptic transmission [57,69]. It cannot be ruled out, however, that a very large potentiation exclusively at the fibers that have undergone conditioning stimulation may produce the NWR facilitation observed during the experiment, since the conditioning stimulation affected an area within the territory of the nerve stimulated to elicit the NWR.

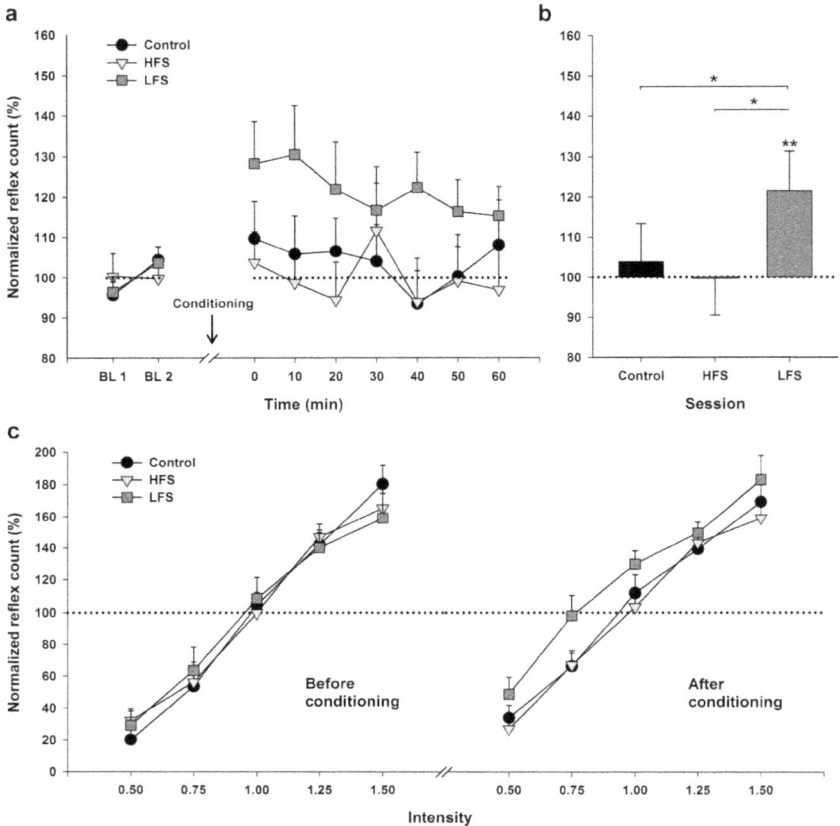

Fig. 3.2. Reflex count. **a** Time course of normalized reflex count before and after conditioning electrical stimulation. **b** Mean values of post-conditioning changes of reflex count across time, in the $0 - 60$ min interval; asterisks on top of the bars indicate significant differences (** $p < 0.01$) on the contrast analysis between pre- and post-conditioning values in each session; asterisks between bars indicate significant differences (* $p < 0.05$) on post-hoc Student-Newman-Keuls tests following RM ANOVA between sessions. **c** Stimulus-response functions for reflex count, before and after conditioning. Dotted lines indicate mean level of baseline period. Mean ± SEM values across 13 volunteers are shown.

3.1.2   Conditioning paradigm

In the past, primarily HFS was shown to evoke LTP in the spinal cord [35,48,56]. Moreover, perceptual correlates of spinal long-term potentiation (a particular type of central sensitization) were obtained in experiments involving human participants [38,41,43]. In SP IV, the HFS paradigm was tested on healthy volunteers using the same electrode and stimulation parameters as in Study III in order to induce central sensitization, the only difference being the site of application (forearm vs. dorsum of the foot). The results in SP IV showed that HFS is capable of producing central sensitization, and that it can be measured not only using subjective behavioral correlates, but also through objective electrophysiological measures like event-related potentials.

LFS, on the other hand, had previously been used mainly for eliciting long-term depression (LTD) of synaptic transmission in the hippocampus [18,52], although this paradigm was later shown to cause a similar effect in spinal cord synapses [60]. More recently, however, sustained LFS has successfully been used as a model for electrically-evoked pain and hyperalgesia in human skin [12,44]. Moreover, it has been shown that a low-frequency afferent barrage at C-fiber intensity, (similar to those produced in inflammation or ectopic beats from neuromas) could also induce LTP at superficial [17,32] and deep [29] dorsal horn neurons.

In line with these observations, the results in Study III showed that LFS at sufficient intensity activated thin nociceptive afferents, including C fibers as judged by the increase in cutaneous blood flow [19,34,62]. The NWR latencies reported there (~ 80 ms) also provide evidence for a spinal mechanism involving Aδ- and C-fiber pathways [1]. Therefore, the findings in Study III support a model of central sensitization where low-frequency activation of primary afferents could induce heterosynaptic activity-dependent amplification in nociceptive processing, most likely due to changes in synaptic transmission within the dorsal horn.

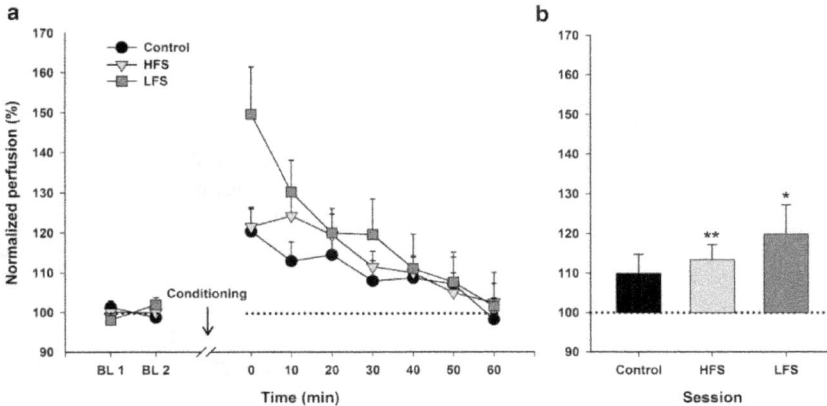

Fig. 3.3. Vascular response. **a** Time course of normalized perfusion before and after conditioning electrical stimulation. **b** Mean values of post-conditioning changes of blood flow across time, in the 0 – 60 min interval; asterisks on top of the bars indicate significant differences (* $p < 0.05$, ** $p < 0.01$) on the contrast analysis between pre- and post-conditioning values. Dotted lines indicate mean level of baseline period. Mean ± SEM values across 13 volunteers are shown.

A different situation was observed after HFS; relatively high pain intensity scores were detected during conditioning stimulation (probably due to an affective response triggered by the stimulation), but the vasodilatation was not significantly different from the Control session and significantly smaller compared to that evoked by LFS, and no modulation of the reflex responses or perceptual measurements was detected. One factor that could influence the outcome of the experiment is the conditioning site; the sensitivity on the dorsum of the foot is probably lower than on the forearm [47], which is the site chosen for most of the trials involving conditioning electrical stimulation and central sensitization performed in humans until now. Another possible explanation for those observations could be that descending inhibition was triggered after high-frequency conditioning stimulation. Although many studies have successfully induced long-lasting facilitation using HFS, others have shown that strong nociceptive input may trigger enhanced descending inhibition, which may overshadow the quantification of long-term facilitation [21,22,61].

This disparity might also be linked to the parameters of the input to the spinal cord (e.g. stimulation intensity and frequency, electrode location and configuration) and the relationships between them, which could explain the different effects observed when using a similar model. Since potentiation and depression of synaptic transmission have been elicited with a variety of stimulation paradigms, it has been suggested that the thresholds for induction of these mechanisms are narrowly tuned and slight changes in experimental conditions can influence the occurrence and polarity of the resulting phenomena [59].

3.2 CAPSAICIN MODEL

The injection of capsaicin provides a unique model to study the mechanisms of central sensitization in humans via a neurogenic inflammation, i.e. capsaicin activates the nerve via the TRPV1 receptor, causing strong firing at the central synapse but also depletion of vasoactive agents in the periphery following antidromic activity. Thus, it resembles the effects of an actual nerve injury (e.g., hyperalgesia, allodynia, enlargement of receptive fields) without any evident tissue damage [40,49,70,73]. Moreover, such effects are evident just seconds after the administration of the substance and may last up to a couple of hours, depending on the delivery method, the dosage and the site of application [47,68].

Variations in RRF were previously shown to reflect changes in central processing of nociceptive activity, for instance after repetitive painful stimulation [5], increased excitability in the nociceptive system (SP III) and alterations in descending control [3]. Since the responses through the reflex pathways are facilitated by sensitization, and that this phenomenon is depending on the site of injury and the degree of descending control [28], it is hypothesized that the descending modulation may affect the RRF control following strong nociceptive input. In this regard, patients with complete spinal cord injury constitute the best human experimental model to test these mechanisms.

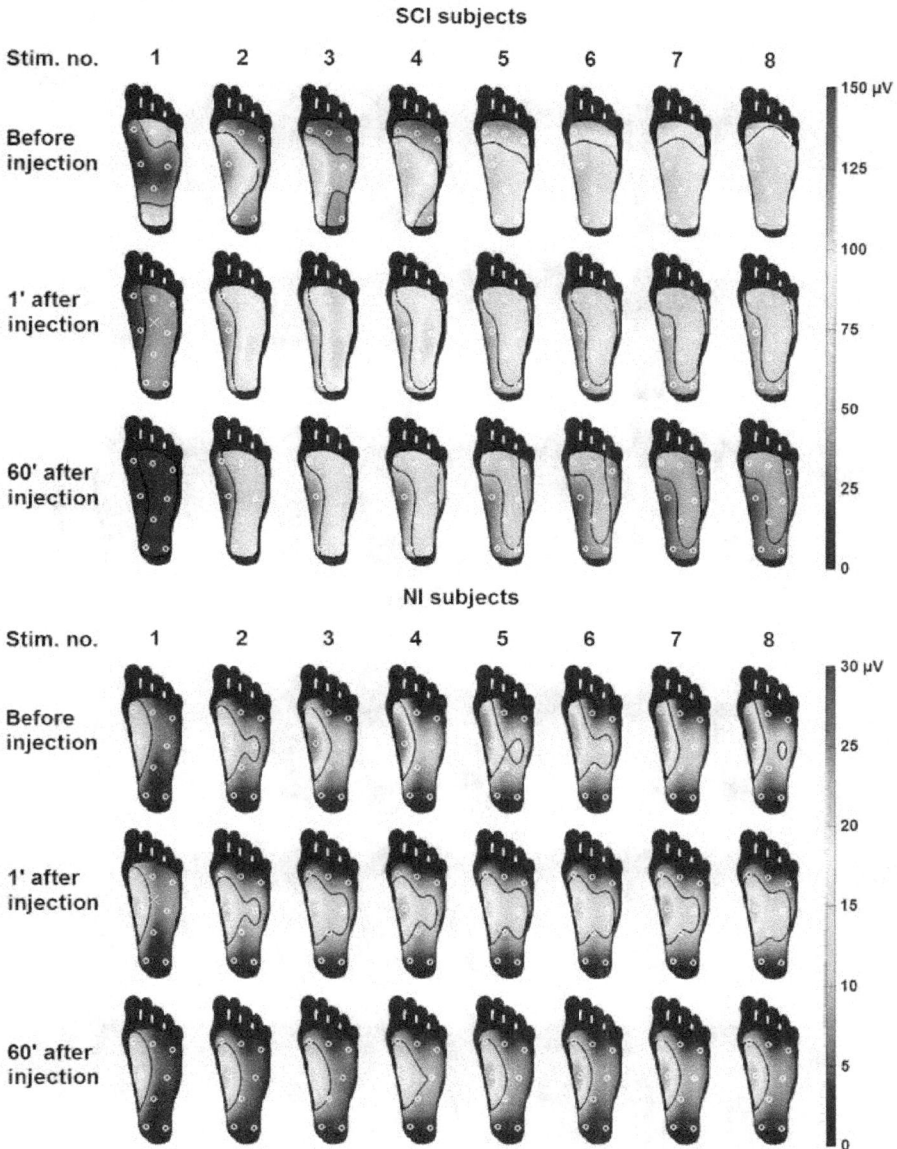

Fig. 3.4. Reflex sensitivity maps for each of the eight stimuli in the train applied to SCI and NI volunteers before, 1 min and 60 min after the capsaicin injection. The black line delimits the RRF sensitivity area. The white cross marks the injection site. Mean RMS amplitudes across all volunteers are shown.

Human Models of Central Sensitization

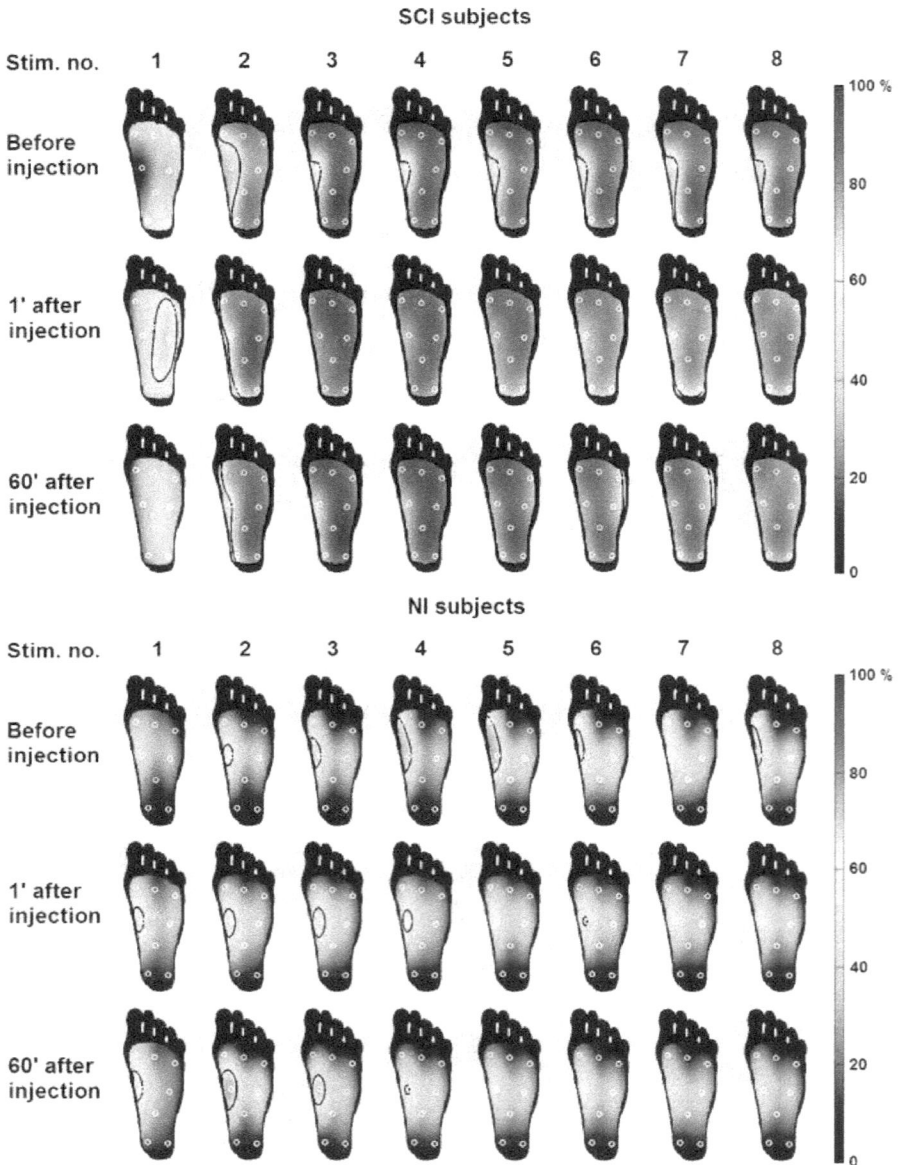

Fig. 3.5. Reflex probability maps for each of the eight stimuli in the train applied to SCI and NI volunteers before, 1 min and 60 min after the capsaicin injection. The black line delimits the RRF probability area. The white cross marks the injection site. Mean probability of occurrence across all volunteers are shown.

In Study IV, the NWR and the RRF were used to investigate the role of descending control on temporal summation and central sensitization (as elicited by capsaicin injection) in humans. Fifteen volunteers with complete spinal cord injury (SCI) and fourteen non-injured (NI) volunteers participated in a single experimental session, where the RRF were assessed before, 1 min after and 60 min after intramuscular injection of capsaicin used to induce central sensitization. In order to elicit temporal summation of the NWR, repeated electrical stimulation was applied on eight sites on the foot sole, and EMG responses were recorded. RRF sensitivity and probability maps were obtained from the EMG using two-dimensional interpolation, and RRF derived measures (area, volume, average probability) were calculated from these maps. The results showed that RRF measures were significantly larger in SCI volunteers compared to NI volunteers, especially during temporal summation of the NWR. Moreover, both groups presented expansion of the RRF immediately after capsaicin injection, as reflected in the enlargement of RRF sensitivity areas and the increase in RRF probability averages (fig. 3.4 and 3.5).

3.2.1   Differences in RRF assessment between SCI and NI volunteers

The NWR has been extensively used to investigate differences in spinal nociception between SCI and NI volunteers, often related to the influence of supraspinal control [31]. After spinal cord transection, the NWR becomes larger and turns into a stereotyped flexor pattern with flexion of all joints [16,25]. Moreover, in accordance with previous evidence gathered from animal experiments [65], the RRF expands dramatically, most likely due to impaired descending control and / or hyperexcitability of spinal neurons [3,24,63].

In Study IV, NWR thresholds were higher in SCI volunteers compared to NI volunteers regardless of the stimulation site, in agreement with previous studies [3,30]. Moreover, the arch of the foot presented the highest thresholds in SCI volunteers and the lowest thresholds in NI volunteers. The RRF also showed the same difference, in which the topography of the sensitivity and probability maps displayed a striking contrast. The most sensitive area in NI volunteers is the arch of the foot, whereas this is completely reversed for SCI volunteers (fig 3.6) The sensitivity of the RRF is shaped by excitatory and inhibitory spinal neuronal circuits under supraspinal influence, among other factors [64]. In chronic SCI volunteers, the loss of descending control and appropriate peripheral input causes a predominance of inhibitory influences on the spinal circuitry [14], that eventually leads to abnormal RRF configurations.

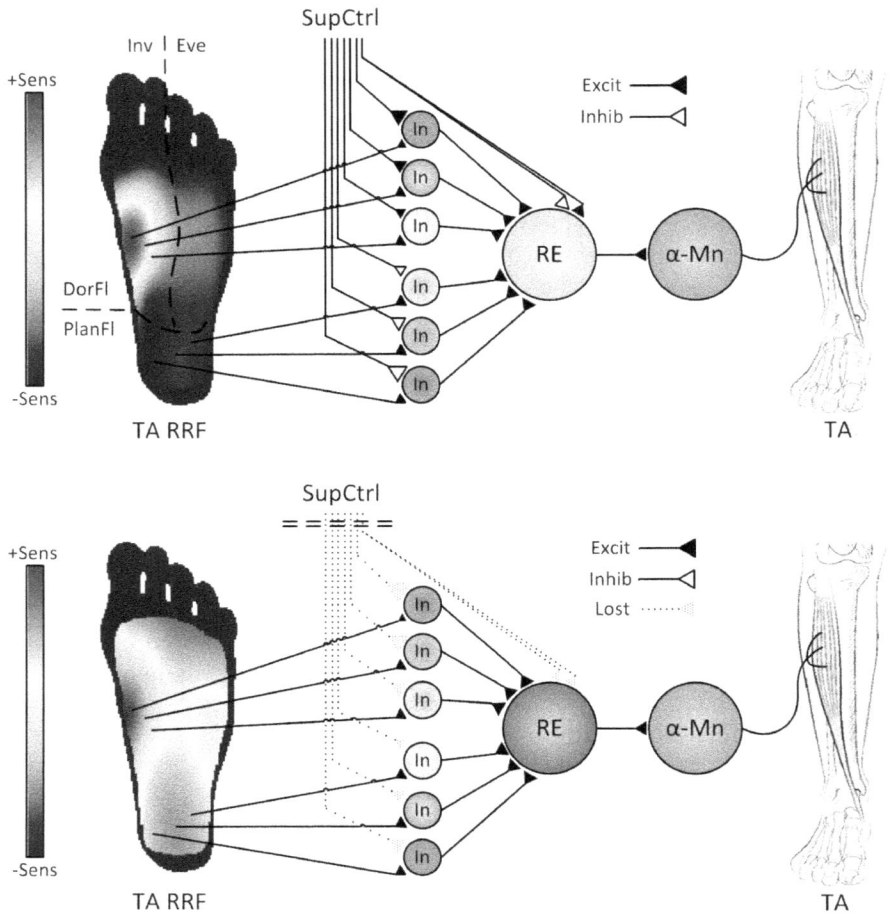

Fig. 3.6. (Left) Functional organization of the nociceptive withdrawal reflex (NWR) pathways. In healthy volunteers, the reflex receptive field (RRF) of the tibialis anterior (TA) muscle is characterized by a high sensitivity (+Sens) in the medial, distal region, resulting in inversion (Inv) and dorsiflexion (DorFl) of the foot when these sites are activated. Functionally antagonist muscles (not shown) have RRF that evoke plantarflexion (PlanFl) or eversion (Eve). The RRF is likely shaped by excitatory (Excit) and inhibitory (Inhib) descending control input coming from supraspinal structures (SupCtrl). SupCtrl may act presynaptically (not shown) or postsynaptically on one or more interneurons (In) and on reflex encoders (RE) in the NWR pathways modifying their excitability (color-coded similarly to RRF), adjusting the weight of afferent information from nociceptive input. The net output is translated by α-motoneurons (α-Mn) into efferent signals that evoke a proper contraction in the target muscle. **b** After an injury to the spinal cord, SupCtrl is partially or totally lost, so In that were subjected to tonic inhibitory descending signals increase their excitability due to disinhibition and vice versa, resulting in abnormal RRF maps.

A marked decrease was noted in NWR amplitudes 60 min after the capsaicin injection compared to the baseline measurements, reflected by the RRF sensitivity volume. This effect could also be noticed right after the capsaicin injection, but the decrease in NWR amplitude was compensated by the enlargement of the RRF sensitivity areas, and therefore the resulting RRF sensitivity volume did not present a significant variation in relation to baseline. Similar phenomena have been described before, in relation to the strong habituation to electrical stimulation that SCI volunteers exhibit during NWR experiments [3], or in locomotion experiments [15], probably indicating a diminished capacity to recruit flexor motoneurons. Antispasticity medication interacting with the GABA system can also be responsible for decreased excitability, yielding smaller reflexes in general [3]. However, it was unlikely to cause this trend within the time course of the experiment (the measurements at every time point should be affected equally), together with the fact that the values in SCI were still significantly larger than in NI volunteers. Interestingly, RRF sensitivity areas and RRF probability measures did not exhibit this behavior, from which it can be hypothesized that these methods might be more robust against habituation.

3.2.2   Central sensitization effects on RRF

In Study IV, an expansion of the RRF sensitivity areas was observed immediately after the capsaicin injection in comparison to baseline measurements. As expected, this effect fades over time, since 60 min after the injection the RRF sensitivity areas were significantly smaller compared to the areas assessed right after the injection; however, at this point they were still larger than the areas elicited before the injection. This RRF behavior is a generalization of previous experiments showing facilitation of the NWR after topical application of capsaicin in humans [2,26]. Moreover, previous findings in animals following chemical irritation showed widespread reflex facilitation distal the knee joint in decerebrated, spinal animals, whereas the facilitation was restricted to specific sites on the sole and ankle in spinally intact animals [28], indicating a supraspinal control on the spinal networks that may be involved in sensitization [1,37,42,46,61], in agreement with the hypothesis of this study (fig. 3.7).

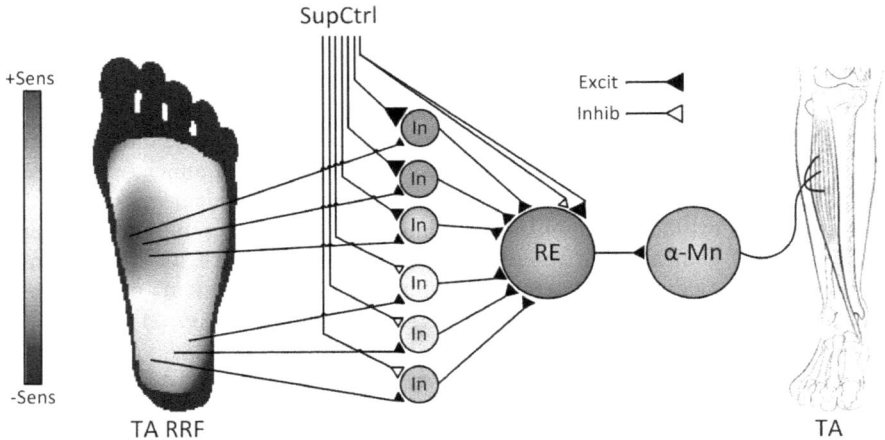

Fig. 3.7. During heterosynaptic sensitization, increased afferent activity (usually through nociceptive C fibres) causes the In and/or RE neurons to become hyperexcitable. Additionally, SupCtrl may increase facilitation and/or decrease inhibition, and as a consequence the RRF is enlarged.

A similar expansion can be observed for RRF probability averages quantified after the 1st stimulus in the train: the average probability of occurrence of a NWR is significantly larger right after the capsaicin injection, although it is not the case for the remaining stimuli in the train. Moreover, such effect was not observed at all for RRF probability areas, which encompass the higher range of probabilities (60-100% occurrence). Together, these results indicated that it is easier to modulate the occurrence of the NWR in the lower range of probabilities (below 50%); it is likely that the descending inhibitory mechanisms play a more important role above this threshold, as can be observed by the much larger RRF probability areas and averages in SCI compared to NI volunteers during temporal summation. Finally, it is worth mentioning that RRF sensitivity and probability maps exhibit complementary (rather than redundant) information, as it is also observed in Study II and III. This could be an indication that the size and occurrence of the NWR are regulated by different neural mechanisms, which in time are strongly modulated by the influence of descending control [71].

3.2.3   Temporal summation effects on RRF

Temporal summation can be described as successive increases in perceived pain intensity to repeated stimuli. As a physiological correlate, the NWR has proven to be a particularly useful tool [6,8,27,67]. In healthy volunteers, it is characterized by a gradual increase in size and duration of the NWR for a few seconds following repetitive stimulation depending on stimulus intensity and frequency [8], after which the NWR size reaches a plateau or even decreases, probably due to descending inhibitory control [10]. Temporal summation can be associated to the early part of the wind-up process that involves summation of excitatory post-synaptic potentials [55], and when assessed by the NWR, is a strong measure of the sensitivity of spinal nociceptive integration, involving N-methyl-D-aspartate (NMDA) receptor pathways that are not detected by single stimuli [7]. Moreover, the NMDA pathways are likely involved in the regulation of descending control [13] and the induction and maintenance of central sensitization [72].

The findings in Study IV showed that RRF sensitivity areas were larger in SCI than in NI volunteers, and this effect was clearly more pronounced during temporal summation. These results are consistent with previous experiments showing enlarged RRF in SCI compared to NI volunteers at different stimulation intensities but using a single electrical stimulus [3], as well as gradual enlargement of the RRF in response to repetitive stimulation in NI volunteers [5]. An interesting finding was that electrical stimulation consistently elicited larger RRF probability areas and RRF probability averages in SCI compared to NI volunteers only during temporal summation (2nd to 8th stimuli in the train); the 1st stimuli in the train, however, did not elicit significant differences between groups. Indeed, spinal facilitatory effects (including temporal summation and central sensitization) can be masked or even completely overridden by supraspinal inhibitory processes triggered after high intensity stimulation [10,23], as demonstrated by the differential effect between NI and SCI volunteers.

REFERENCES

[1]   Andersen OK. Studies of the organization of the human nociceptive withdrawal reflex. Focus on sensory convergence and stimulation site dependency. Acta Physiol 2007;189:1-35.

[2]   Andersen OK, Felsby S, Nicolaisen L, Bjerring P, Jensen TS, Arendt-Nielsen L. The effect of Ketamine on stimulation of primary and secondary hyperalgesic areas induced by capsaicin - A double-blind, placebo-controlled, human experimental study. Pain 1996;66:51-62.

[3]   Andersen OK, Finnerup NB, Spaich EG, Jensen TS, Arendt-Nielsen L. Expansion of nociceptive withdrawal reflex receptive fields in spinal cord injured humans. Clin Neurophysiol 2004;115:2798-2810.

[4]   Andersen OK, Jensen LM, Brennum J, Arendt-Nielsen L. Evidence for central summation of C and A[delta] nociceptive activity in man. Pain 1994;59:273-280.

[5]   Andersen OK, Spaich EG, Madeleine P, Arendt-Nielsen L. Gradual enlargement of human withdrawal reflex receptive fields following repetitive painful stimulation. Brain Res 2005;1042:194-204.

[6]   Arendt-Nielsen L, Brennum J, Sindrup S, Bak P. Electrophysiological and psychophysical quantification of temporal summation in the human nociceptive system. Eur J Appl Physiol Occup Physiol 1994;68:266-273.

[7]   Arendt-Nielsen L, Petersen-Felix S, Fischer M, Bak P, Bjerring P, Zbinden AM. The effect of N-methyl-D-aspartate antagonist (ketamine) on single and repeated nociceptive stimuli: A placebo-controlled experimental human study. Anesth Analg 1995;81:63-68.

[8]   Arendt-Nielsen L, Sonnenborg FA, Andersen OK. Facilitation of the withdrawal reflex by repeated transcutaneous electrical stimulation: An experimental study on central integration in humans. Eur J Appl Physiol Occup Physiol 2000;81:165-173.

[9]   Bailey CH, Giustetto M, Huang Y-, Hawkins RD, Kandel ER. Is heterosynaptic modulation essential for stabilizing Hebbian plasticity and memory? Nat Rev Neurosci 2000;1:11-20.

[10]   Bajaj P, Arendt-Nielsen L, Andersen OK. Facilitation and inhibition of withdrawal reflexes following repetitive stimulation: electro- and psychophysiological evidence for activation of noxious inhibitory controls in humans. Eur J Pain 2005;9:25-31.

[11]   Brain SD, Williams TJ. Substance P regulates the vasodilator activity of calcitonin gene-related peptide. Nature 1988;335:73-75.

[12]   Chizh BA, Gö□hring M, Trö□ster A, Quartey GK, Schmelz M, Koppert W. Effects of oral pregabalin and aprepitant on pain and central sensitization in the electrical hyperalgesia model in human volunteers. Br J Anaesth 2007;98:246-254.

[13]   Clarke RW, Eves S, Harris J, Peachey JE, Stuart E. Interactions between cutaneous afferent inputs to a withdrawal reflex in the decerebrated rabbit and their control by descending and segmental systems. Neuroscience 2002;112:555-571.

[14]   Dietz V. Behavior of spinal neurons deprived of supraspinal input. Nat Rev Neurol 2010;6:167-174.

[15]   Dietz V, Müller R. Degradation of neuronal function following a spinal cord injury: mechanisms and countermeasures. Brain 2004;127:2221-2231.

[16]   Dimitrijević MR, Nathan PW. Studies of spasticity in man: 3. Analysis of reflex activity evoked by noxious cutaneous stimulation. Brain 1968;91:349-368.

[17]   Drdla R, Sandkühler J. Long-term potentiation at C-fibre synapses by low-level presynaptic activity in vivo. Mol Pain 2008;4.

[18] Dudek SM, Bear MF. Homosynaptic long-term depression in area CA1 of hippocampus and effects of N-methyl-D-aspartate receptor blockade. Proc Natl Acad Sci USA 1992;89:4363-4367.

[19] Dusch M, Schley M, Rukwied R, Schmelz M. Rapid flare development evoked by current frequency-dependent stimulation analyzed by full-field laser perfusion imaging. Neuroreport 2007;18:1101-1105.

[20] Geber C, Fondel R, Kra☐mer HH, Rolke R, Treede R-, Sommer C, Birklein F. Psychophysics, Flare, and Neurosecretory Function in Human Pain Models: Capsaicin Versus Electrically Evoked Pain. J.Pain 2007;8:503-514.

[21] Gjerstad J, Tjølsen A, Hole K. Induction of long-term potentiation of single wide dynamic range neurones in the dorsal horn is inhibited by descending pathways. Pain 2001;91:263-268.

[22] Gjerstad J, Tjølsen A, Svendsen F, Hole K. Inhibition of spinal nociceptive responses after intramuscular injection of capsaicin involves activation of noradrenergic and opioid systems. Brain Res 2000;859:132-136.

[23] Gozariu M, Bragard D, Willer J-, Le Bars D. Temporal summation of C-fiber afferent inputs: Competition between facilitatory and inhibitory effects on C-fiber reflex in the rat. J Neurophysiol 1997;78:3165-3179.

[24] Grimby L. Normal plantar response: integration of flexor and extensor reflex components. J Neurol Neurosurg Psychiatr 1963;26:39-50.

[25] Grimby L. Pathological plantar response: disturbances of the normal integration of flexor and extensor reflex components. J Neurol Neurosurg Psychiatr 1963;26:314-321.

[26] Grönroos M, Pertovaara A. Capsaicin-induced central facilitation of a nociceptive flexion reflex in humans. Neurosci Lett 1993;159:215-218.

[27] Guirimand F, Dupont X, Brasseur L, Chauvin M, Bouhassira D. The effects of ketamine on the temporal summation (wind-up) of the R(III) nociceptive flexion reflex and pain in humans. Anesth Analg 2000;90:408-414.

[28] Harris J, Clarke RW. Organisation of sensitisation of hind limb withdrawal reflexes from acute noxious stimuli in the rabbit. J Physiol (Lond) 2003;546:251-265.

[29] Haugan F, Wibrand K, Fiskå A, Bramham CR, Tjølsen A. Stability of long term facilitation and expression of zif268 and Arc in the spinal cord dorsal horn is modulated by conditioning stimulation within the physiological frequency range of primary afferent fibers. Neurosci 2008;154:1568-1575.

[30] Hiersemenzel L-, Curt A, Dietz V. From spinal shock to spasticity: Neuronal adaptations to a spinal cord injury. Neurology 2000;54:1574-1582.

[31] Hornby TG, Rymer WZ, Benz EN, Schmit BD. Windup of flexion reflexes in chronic human spinal cord injury: A marker for neuronal plateau potentials? J Neurophysiol 2003;89:416-426.

[32] Ikeda H, Stark J, Fischer H, Wagner M, Drdla R, Jäger T, Sandkühler J. Synaptic amplifier of inflammatory pain in the spinal dorsal horn. Science 2006;312:1659-1662.

[33] Inui K, Tran TD, Hoshiyama M, Kakigi R. Preferential stimulation of Aδ fibers by intra-epidermal needle electrode in humans. Pain 2002;96:247-252.

[34] Jänig W, Lisney SJW. Small diameter myelinated afferents produce vasodilatation but not plasma extravasation in rat skin. J Physiol (Lond) 1989;415:477-486.

[35] Ji RR, Kohno T, Moore KA, Woolf CJ. Central sensitization and LTP: do pain and memory share similar mechanisms? Trends Neurosci 2003;26:696-705.

[36] Kidd BL, Urban LA. Mechanisms of inflammatory pain. Br J Anaesth 2001;87:3-11.

[37] Klauenberg S, Maier C, Assion H-, Hoffmann A, Krumova EK, Magerl W, Scherens A, Treede R-, Juckel G. Depression and changed pain perception: Hints for a central disinhibition mechanism. Pain 2008;140:332-343.

[38] Klein T, Magerl W, Hopf HC, Sandkühler J, Treede R-. Perceptual Correlates of Nociceptive Long-Term Potentiation and Long-Term Depression in Humans. J Neurosci 2004;24:964-971.

[39] Klein T, Magerl W, Nickel U, Hopf H-, Sandkühler J, Treede R-. Effects of the NMDA-receptor antagonist ketamine on perceptual correlates of long-term potentiation within the nociceptive system. Neuropharmacol 2007;52:655-661.

[40] Klein T, Magerl W, Rolke R, Treede R-. Human surrogate models of neuropathic pain. Pain 2005;115:227-233.

[41] Klein T, Magerl W, Treede R-. Perceptual Correlate of Nociceptive Long-Term Potentiation (LTP) in Humans Shares the Time Course of Early-LTP. J Neurophysiol 2006;96:3551-3555.

[42] Klein T, Magerl W, Treede R-. Forget about your chronic pain. Pain 2007;132:16-17.

[43] Klein T, Stahn S, Magerl W, Treede R-. The role of heterosynaptic facilitation in long-term potentiation (LTP) of human pain sensation. Pain 2008;139:507-519.

[44] Koppert W, Dern SK, Sittl R, Albrecht S, Schu□ttler J, Schmelz M. A new model of electrically evoked pain and hyperalgesia in human skin: The

effects of intravenous alfentanil, S(+)-ketamine, and lidocaine. Anesthesiol 2001;95:395-402.

[45] Latremoliere A, Woolf CJ. Central Sensitization: A Generator of Pain Hypersensitivity by Central Neural Plasticity. J Pain 2009;10:895-926.

[46] Lemon RN, Griffiths J. Comparing the function of the corticospinal system in different species: Organizational differences for motor specialization? Muscle Nerve 2005;32:261-279.

[47] Liu M, Max MB, Robinovitz E, Gracely RH, Bennett GJ. The human capsaicin model of allodynia and hyperalgesia: Sources of variability and methods for reduction. J Pain Symptom Manag 1998;16:10-20.

[48] Liu X-, Sandkühler J. Characterization of long-term potentiation of C-fiber-evoked potentials in spinal dorsal horn of adult rat: Essential role of NK1 and NK2 receptors. J Neurophysiol 1997;78:1973-1982.

[49] Magerl W, Fuchs PN, Meyer RA, Treede R-. Roles of capsaicin-insensitive nociceptors in cutaneous pain and secondary hyperalgesia. Brain 2001;124:1754-1764.

[50] Magerl W, Wilk SH, Treede R-. Secondary hyperalgesia and perceptual wind-up following intradermal injection of capsaicin in humans. Pain 1998;74:257-268.

[51] McCarthy PW, Lawson SN. Cell type and conduction velocity of rat primary sensory neurons with substance P-like immunoreactivity. Neuroscience 1989;28:745-753.

[52] Mulkey RM, Malenka RC. Mechanisms underlying induction of homosynaptic long-term depression in area CA1 of the hippocampus. Neuron 1992;9:967-975.

[53] Nilsson H-, Schouenborg J. Differential inhibitory effect on human nociceptive skin senses induced by local stimulation of thin cutaneous fibers. Pain 1999;80:103-112.

[54] Prescott SA. Interactions between depression and facilitation within neural networks: Updating the dual-process theory of plasticity. Learn Memory 1998;5:446-466.

[55] Randic M. Plasticity of excitatory synaptic transmission in the spinal cord dorsal horn. PROG BRAIN RES 1996;113:463-506.

[56] Randić M, Jiang MC, Cerne R. Long-term potentiation and long-term depression of primary afferent neurotransmission in the rat spinal cord. J Neurosci 1993;13:5228-5241.

[57] Rygh LJ, Tjølsen A, Hole K, Svendsen F. Cellular memory in spinal nociceptive circuitry. Scand J Psychol 2002;43:153-159.

[58] Sandkühler J. Learning and memory in pain pathways. Pain 2000;88:113-118.

[59] Sandkühler J. Understanding LTP in pain pathways. Mol Pain 2007;3.

[60] Sandkühler J, Chen JG, Cheng G, Randić M. Low-frequency stimulation of afferent Aδ-fibers induces long-term depression at primary afferent synapses with substantia gelatinosa neurons in the rat. J Neurosci 1997;17:6483-6491.

[61] Sandkühler J, Liu X. Induction of long-term potentiation at spinal synapses by noxious stimulation or nerve injury. Eur J Neurosci 1998;10:2476-2480.

[62] Schmelz M, Michael K, Weidner C, Schmidt R, Torebjörk HE, Handwerker HO. Which nerve fibers mediate the axon reflex flare in human skin? Neuroreport 2000;11:645-648.

[63] Schmit BD, Hornby TG, Tysseling-Mattiace VM, Benz EN. Absence of Local Sign Withdrawal in Chronic Human Spinal Cord Injury. J Neurophysiol 2003;90:3232-3241.

[64] Schouenborg J. Modular organisation and spinal somatosensory imprinting. Brain Res Rev 2002;40:80-91.

[65] Schouenborg J, Holmberg H, Weng HR. Functional organization of the nociceptive withdrawal reflexes. Exp Brain Res 1992;90:469-478.

[66] Schouenborg J, Sjolund BH. Activity evoked by A- and C-afferent fibers in rat dorsal horn neurons and its relation to a flexion reflex. J Neurophysiol 1983;50:1108-1121.

[67] Serrao M, Rossi P, Sandrini G, Parisi L, Amabile GA, Nappi G, Pierelli F. Effects of diffuse noxious inhibitory controls on temporal summation of the RIII reflex in humans. Pain 2004;112:353-360.

[68] Simone DA, Baumann TK, LaMotte RH. Dose-dependent pain and mechanical hyperalgesia in humans after intradermal injection of capsaicin. Pain 1989;38:99-107.

[69] Svendsen F, Tjølsen A, Gjerstad J, Hole K. Long term potentiation of single WDR neurons in spinalized rats. Brain Res 1999;816:487-492.

[70] Treede R-, Meyer RA, Raja SN, Campbell JN. Peripheral and central mechanisms of cutaneous hyperalgesia. Prog Neurobiol 1992;38:397-421.

[71] Willis Jr. WD. Anatomy and physiology of descending control of nociceptive responses of dorsal horn neurons: Comprehensive review. Prog Brain Res 1988;77:1-29.

[72] Woolf CJ, Thompson SWN. The induction and maintenance of central sensitization is dependent on N-methyl-D-aspartic acid receptor activation;

implications for the treatment of post-injury pain hypersensitivity states. Pain 1991;44:293-299.

[73] Ziegler EA, Magerl W, Meyer RA, Treede R-. Secondary hyperalgesia to punctate mechanical stimuli. Central sensitization to A-fibre nociceptor input. Brain 1999;122:2245-2257.

Chapter 4.

Conclusion

The results from Study I and II showed that the NWR and the RRF are robust and reliable tools in experimental and clinical pain research. In particular, the parameters that can be derived from RRF sensitivity and probability maps (areas, volumes, averages) are not significantly affected by factors like habituation to single and repeated electrical stimulation and small variations in stimulation intensities. With that in mind, the NWR and the RRF were used in Study III and IV in the assessment of human surrogate models of central sensitization.

In Study III, a central sensitization model using noxious conditioning electrical stimulation was tested. Persistent facilitation of the NWR was observed following low-frequency stimulation, probably mediated by thin primary afferents. Although supraspinal interactions cannot be completely ruled out, the most likely neuronal mechanism that could explain these observations involves heterosynaptic interactions within the spinal dorsal horn. In Study IV, an intramuscular injection of capsaicin was used to induce central sensitization in both healthy and complete spinal cord injured volunteers. The results showed that RRF were significantly modulated after the injection in both groups, and that this modulation was under strong influence of descending control, as demonstrated by the differential effects in size and shape of the RRF observed in spinal cord injured volunteers compared to healthy volunteers.

In summary, the studies presented in this thesis have hopefully contributed to a better understanding of human models of central sensitization, as well as the establishment of the NWR and RRF as viable alternatives for objective assessment of central changes in spinal nociception.

4.1 FUTURE PERSPECTIVES

Several methodological aspects of the NWR and RRF assessment can still be addresseed in order to further improve the robustness and reliability of these tools towards widespread clinical application. Among these it is worth mentioning further standardization of NWR and RRF recordings, new methods to reduce cross-talk in EMG signals based (for instance based on double differential recordings) and technological improvements on electrode in order to minimize differences arising from skin thickness / impedance on the sole of the foot.

Regarding human models of central sensitization, the main goal for future research should focus on the investigation of which are the optimal experimental conditions (e.g., type and dose of algogenic substances, conditioning electrical stimulation parameters) that are able to elicit stable, reproducible effects over time, while still closely resembling the mechanisms behind clinical pathophysiological conditions. Once that is accomplished, further research can be directed into testing new drugs or alternative methods to modulate these effects, in order to develop better alternatives for pain relief.

About the author

José A. Biurrun Manresa was born in Salta, Argentina in 1982. He graduated as Bioengineer in 2007 at the Faculty of Engineering of the National University of Entre Ríos (Argentina). In 2011, he obtained his Ph.D. degree from the Department of Health Science and Technology at Aalborg University (Denmark), where he is currently employed as Assistant Professor. His main areas of research are biomedical signal processing with focus on EMG and EEG, and the neurophysiology of spinal and supraspinal nociception.

www.ingramcontent.com/pod-product-compliance
Lightning Source LLC
Chambersburg PA
CBHW061840220326
41599CB00027B/5349